U.S.NRC

United States Nuclear Regulatory Commission

Protecting People and the Environment

NUREG-1022, Rev. 3

I0482662

Event Report Guidelines 10 CFR 50.72 and 50.73

Final Report

Office of Nuclear Reactor Regulation

AVAILABILITY OF REFERENCE MATERIALS
IN NRC PUBLICATIONS

United States Nuclear Regulatory Commission

Protecting People and the Environment

NUREG-1022, Rev. 3

Event Report Guidelines 10 CFR 50.72 and 50.73

Final Report

Manuscript Completed: January 2013
Date Published: January 2013

Prepared by:
Aron Lewin

Office of Nuclear Reactor Regulation

ABSTRACT

NUREG-1022, "Event Reporting Guidelines: 10 CFR 50.72 and 50.73," contains guidelines that the staff of the U.S. Nuclear Regulatory Commission (NRC) considers acceptable for use in meeting the requirements of Title 10 of the Code of Federal Regulations (10 CFR) 50.72 and 50.73. Several identified reporting issues could not be quickly resolved given certain ambiguities in NUREG-1022, Revision 2 guidance. In developing Revision 3 to NUREG-1022, the NRC held numerous public and internal meetings to solicit stakeholder inputs and feedback. In resolving the ambiguities, the NRC considered the provisions of the rule itself, the associated statements of consideration, and other available guidance in that hierarchal order. Revision 3 to NUREG-1022 revises the event reporting guidelines in order to provide clearer guidance.

CONTENTS

EXECUTIVE SUMMARY

Two of the many elements contributing to the safety of nuclear power are emergency response and the feedback from operating experience into plant operations. These are achieved partly by the licensee event reporting requirements of Title 10 of the *Code of Federal Regulations* (10 CFR) 50.72, "Immediate Notification Requirements for Operating Nuclear Power Reactors," and 10 CFR 50.73, "Licensee Event Report System." In 10 CFR 50.72, the U.S. Nuclear Regulatory Commission (NRC) provides for immediate notification requirements through the emergency notification system, and in 10 CFR 50.73 provides for 60-day written licensee event reports.

The NRC staff uses the information reported under 10 CFR 50.72 and 10 CFR 50.73 in responding to emergencies, monitoring ongoing events, confirming licensing bases, studying potentially generic safety problems, assessing trends and patterns of operational experience, monitoring performance, identifying precursors of more significant events, and providing operational experience to the industry.

NUREG-1022 contains guidelines that the staff of the NRC considers acceptable for use in meeting the requirements of 10 CFR 50.72 and 10 CFR 50.73. Several identified reporting issues could not be quickly resolved given certain ambiguities in the guidance in Revision 2 of NUREG-1022. In developing Revision 3 to NUREG-1022, the NRC held numerous public and internal meetings to solicit stakeholder input and feedback. In resolving the ambiguities, the NRC considered the provisions of the rule itself, the associated statements of consideration, and other available guidance in that hierarchal order. Revision 3 to NUREG-1022 revises the event reporting guidelines to provide clearer guidance.

ABBREVIATIONS

ADAMS	Agencywide Documents Access and Management System
AFW	auxiliary feedwater
ALI	annual limits on intake
ANS	American Nuclear Society
ASME	American Society of Mechanical Engineers
BWR	boiling-water reactor
CFR	*Code of Federal Regulations*
CRDM	control rod drive mechanism
CREFS	Control Room Emergency Filtration System
DAC	derived air concentration
ECCS	emergency core cooling system
EDG	emergency diesel generator
EIIS	Energy Industry Identification System
ENS	emergency notification system
EO	emergency officer
EOF	emergency operations facility
EPIX	Equipment Performance and Information Exchange
EPA	Environmental Protection Agency (U.S.)
EPZ	emergency planning zone
ERDS	Emergency Response Data System
ERF	emergency response facility
ESW	emergency service water
FEMA	Federal Emergency Management Agency
FSAR	final safety analysis report
HOO	headquarters operations officer
HPCI	high-pressure coolant injection
HPI	high-pressure injection
HPN	health physics network
HPSI	high-pressure safety injection
IEEE	Institute of Electrical and Electronics Engineers
IN	information notice
kV	kilovolt
LCO	limiting condition for operation
LER	licensee event report
MSIV	main steam isolation valve
NESP	National Environmental Studies Project
NRC	Nuclear Regulatory Commission (U.S.)

NUMARC   Nuclear Management and Resources Council

OMS   overpressure mitigation system

PWR   pressurized-water reactor

RCIC   reactor core isolation cooling
RCS   reactor coolant system
RDO   regional duty officer
REP   radiological emergency preparedness
RHR   residual heat removal
RIS   regulatory issue summary
RPS   reactor protection system
RWCU   reactor water cleanup

SI   safety injection
SPDS   safety parameter display system
SSC   system, structure, and/or component
STS   Standard Technical Specification(s)

TS   technical specification(s)
TSC   technical support center

VFTP   Ventilation Filter Testing Program

1. INTRODUCTION

This document provides guidance on the reporting requirements of Title 10 of the *Code of Federal Regulations* (10 CFR) 50.72, "Immediate Notification Requirements for Operating Nuclear Power Reactors," and 10 CFR 50.73, Licensee Event Report System." While these reporting requirements range from immediate, 1-hour, 4-hour, and 8-hour telephone notifications to 60-day written reports, covering a broad spectrum of events from emergencies to component level deficiencies, the U.S. Nuclear Regulatory Commission (NRC) wishes to emphasize that reporting requirements should not interfere with ensuring the safe operation of a nuclear power plant. Licensees' immediate attention must always be given to operational safety concerns.

1.1 Background

In 1983, partially in response to lessons from the accident at Three Mile Island Nuclear Station, the NRC revised its immediate notification requirements through the emergency notification system (ENS) in 10 CFR 50.72 and modified and codified its written licensee event report (LER) system requirements in 10 CFR 50.73. The revision to 10 CFR 50.72 and the new 10 CFR 50.73 became effective on January 1, 1984. Together, they specified the types of events and conditions reportable to the NRC for emergency response and identified plant-specific and generic safety issues. They remained in effect, with only minor modifications until early 2001.

In late October 2000, the NRC published substantial amendments to 10 CFR 50.72 and 10 CFR 50.73 in the *Federal Register*, with an effective date in late January 2001.

1.2 Revised Reporting Guidelines

The purpose of this Revision 3 to NUREG-1022 is to incorporate changes to the guidelines for the purpose of clarification. This report supersedes Revision 2 to NUREG-1022, issued in October 2000.

Section 2 clarifies specific areas of 10 CFR 50.72 and 50.73 that are applicable to multiple reporting criteria or that historically appear to be subject to varied interpretations. It covers such diverse subjects as engineering judgment, differences in tenses between the two rules, retraction and voluntary reporting, legal reporting requirements, and human performance issues.

Section 3 contains guidelines on event reporting for specific criteria in both rules using discussions and examples of reported events. To minimize repetition, similar criteria from both rules are addressed together. Section 3.1 addresses general ENS and LER reporting requirements. Section 3.2 addresses specific ENS and LER reporting criteria. It includes a comprehensive discussion of each specific reporting criterion, with illustrative examples and definitions of key terms and phrases. Section 3.3 addresses the requirements for immediate ENS followup notifications during the course of an event.

Section 4 explains ENS communication reporting timeliness and completeness, voluntary notifications, and retractions. Appropriate ENS emergency notification methods are described.

Section 5 provides guidelines on administrative requirements, as well as the preparation, and submittal of LERs. It specifies the information an LER should contain and provides steps to be followed in preparing an LER. It also includes an expanded human performance discussion to achieve ENS and LER content that examines both equipment and human performance.

Table 1 Reportable Events

Declaration of an Emergency Class (See Section 3.1.1 of this report)	
§ 50.72(a)(1)(i) "The declaration of any of the Emergency Classes specified in the licensee's approved Emergency Plan."	
Plant Shutdown Required by Technical Specifications **(See Section 3.2.1 of this report)**	
§ 50.72(b)(2)(i) "The initiation of any nuclear plant shutdown required by the plant's Technical Specifications."	**§ 50.73(a)(2)(i)(A)** "The completion of any nuclear plant shutdown required by the plant's Technical Specifications."
Operation or Condition Prohibited by Technical Specifications **(See Section 3.2.2 of this report)**	
	§ 50.73(a)(2)(i)(B) "Any operation or condition which was prohibited by the plant's Technical Specifications except when: (1) The Technical Specification is administrative in nature; (2) The event consisted solely of a case of a late surveillance test where the oversight was corrected, the test was performed, and the equipment was found to be capable of performing its specified safety functions; or (3) The Technical Specification was revised prior to discovery of the event such that the operation or condition was no longer prohibited at the time of discovery of the event."
Deviation from Technical Specifications Authorized under § 50.54(x) **(See Section 3.2.3 of this report)**	
§ 50.72(b)(1) "... any deviation from the plant's Technical Specifications authorized pursuant to § 50.54(x) of this part."	**§ 50.73(a)(2)(i)(C)** "Any deviation from the plant's Technical Specifications authorized pursuant to § 50.54(x) of this part."
Degraded or Unanalyzed Condition **(See Section 3.2.4 of this report)**	
§ 50.72(b)(3)(ii) "Any event or condition that	**50.73(a)(2)(ii)** "Any event or condition that

Table 1 Reportable Events (continued)

results in:	resulted in:
(A) The condition of the nuclear power plant, including its principal safety barriers, being seriously degraded; or	(A) The condition of the nuclear power plant, including its principal safety barriers, being seriously degraded; or
(B) The nuclear power plant being in an unanalyzed condition that significantly degrades plant safety."	(B) The nuclear power plant being in an unanalyzed condition that significantly degraded plant safety."

| **External Threat or Hampering**
(See Section 3.2.5 of this report) ||

	§ **50.73(a)(2)(iii)** "Any natural phenomenon or other external condition that posed an actual threat to the safety of the nuclear power plant or significantly hampered site personnel in the performance of duties necessary for the safe operation of the nuclear power plant."

| **System Actuation**
(See Section 3.2.6 of this report) ||

§ **50.72(b)(2)(iv)(A)** "Any event that results or should have resulted in emergency core cooling system (ECCS) discharge into the reactor coolant system as a result of a valid signal except when the actuation results from and is part of a pre-planned sequence during testing or reactor operation." § **50.72(b)(2)(iv)(B)** "Any event or condition that results in actuation of the reactor protection system (RPS) when the reactor is critical except when the actuation results from and is part of a pre-planned sequence during testing or reactor operation." § **50.72(b)(3)(iv)(A)** "Any event or condition that results in valid actuation of any of the systems listed in paragraph (b)(3)(iv)(B) of this section, except when the actuation results from and is part of a pre-planned sequence during testing or reactor operation."	§ **50.73(a)(2)(iv)(A)** "Any event or condition that resulted in manual or automatic actuation of any of the systems listed in paragraph (a)(2)(iv)(B) of this section, except when: (1) The actuation resulted from and was part of a pre-planned sequence during testing or reactor operation; or

Table 1 Reportable Events (continued)

	(2) The actuation was invalid and;
	    (i) Occurred while the system was properly removed from service; or
	    (ii) Occurred after the safety function had been already completed."

§ 50.72(b)(3)(iv)(B) "The systems to which the requirements of paragraph (b)(3)(iv)(A) of this section apply are:	**§ 50.73(a)(2)(iv)(B)** "The systems to which the requirements of paragraph (a)(2)(iv)(A) of this section apply are:
(1) Reactor protection system (RPS) including: reactor scram and reactor trip.[5]	(1) Reactor protection system (RPS) including: reactor scram or reactor trip.
(2) General containment isolation signals affecting containment isolation valves in more than one system or multiple main steam isolation valves (MSIVs).	(2) General containment isolation signals affecting containment isolation valves in more than one system or multiple main steam isolation valves (MSIVs).
(3) Emergency core cooling systems (ECCS) for pressurized water reactors (PWRs) including: high-head, intermediate-head, and low-head injection systems and the low pressure injection function of residual (decay) heat removal systems.	(3) Emergency core cooling systems (ECCS) for pressurized water reactors (PWRs) including: high-head, intermediate-head, and low-head injection systems and the low pressure injection function of residual (decay) heat removal systems.
(4) ECCS for boiling water reactors (BWRs) including: high-pressure and low-pressure core spray systems; high-pressure coolant injection system; low pressure injection function of the residual heat removal system.	(4) ECCS for boiling water reactors (BWRs) including: high-pressure and low-pressure core spray systems; high-pressure coolant injection system; low pressure injection function of the residual heat removal system.
(5) BWR reactor core isolation cooling system; isolation condenser system; and feedwater coolant injection system.	(5) BWR reactor core isolation cooling system; isolation condenser system; and feedwater coolant injection system.
(6) PWR auxiliary or emergency feedwater system.	(6) PWR auxiliary or emergency feedwater system.
(7) Containment heat removal and depressurization systems, including containment spray and fan cooler systems.	(7) Containment heat removal and depressurization systems, including containment spray and fan cooler systems.

---
[5] Actuation of the RPS when the reactor is critical is reportable under § 50.72(b)(2)(iv)(B).

Table 1 Reportable Events (continued)

(8) Emergency ac electrical power systems, including: emergency diesel generators (EDGs); hydroelectric facilities used in lieu of EDGs at the Oconee Station; and BWR dedicated Division 3 EDGs."	(8) Emergency ac electrical power systems, including: emergency diesel generators (EDGs); hydroelectric facilities used in lieu of EDGs at the Oconee Station; and BWR dedicated Division 3 EDGs. (9) Emergency service water systems that do not normally run and that serve as ultimate heat sinks."
Event or Condition that Could Have Prevented Fulfillment of a Safety Function **(See Section 3.2.7 of this report)**	
§ 50.72(b)(3)(v) "Any event or condition that at the time of discovery could have prevented the fulfillment of the safety function of structures or systems that are needed to: (A) Shut down the reactor and maintain it in a safe shutdown condition; (B) Remove residual heat; (C) Control the release of radioactive material; or (D) Mitigate the consequences of an accident." **§ 50.72(b)(3)(vi)** "Events covered in paragraph (b)(3)(v) of this section may include one or more procedural errors, equipment failures, and/or discovery of design, analysis, fabrication, construction, and/or procedural inadequacies. However, individual component failures need not be reported pursuant to paragraph (b)(3)(v) of this section if redundant equipment in the same system was operable and available to perform the required safety function."	**§ 50.73(a)(2)(v)** "Any event or condition that could have prevented the fulfillment of the safety function of structures or systems that are needed to: (A) Shut down the reactor and maintain it in a safe shutdown condition; (B) Remove residual heat; (C) Control the release of radioactive material; or (D) Mitigate the consequences of an accident." **§ 50.73(a)(2)(vi)** "Events covered in paragraph (a)(2)(v) of this section may include one or more procedural errors, equipment failures, and/or discovery of design, analysis, fabrication, construction, and/or procedural inadequacies. However, individual component failures need not be reported pursuant to paragraph (a)(2)(v) of this section if redundant equipment in the same system was operable and available to perform the required safety function."
Common Cause Inoperability of Independent Trains or Channels **(See Section 3.2.8 of this report)**	
	§ 50.73(a)(2)(vii) "Any event where a single cause or condition caused at least one independent train or channel to become inoperable in multiple systems or two independent trains or channels to become inoperable in a single system designed to: (A) Shut down the reactor and maintain it in a safe shutdown condition;

Table 1 Reportable Events (continued)

	(B) Remove residual heat; (C) Control the release of radioactive material; or (D) Mitigate the consequences of an accident."
Radioactive Release **(See Section 3.2.9 of this report)**	
	§ 50.73(a)(2)(viii)(A) "Any airborne radioactive release that, when averaged over a time period of 1 hour, resulted in airborne radionuclide concentrations in an unrestricted area that exceeded 20 times the applicable concentration limits specified in appendix B to part 20, table 2, column 1." **§ 50.73(a)(2)(viii)(B)** "Any liquid effluent release that, when averaged over a time period of 1 hour, exceeds 20 times the applicable concentrations specified in appendix B to part 20, table 2, column 2, at the point of entry into the receiving waters (i.e., unrestricted area) for all radionuclides except tritium and dissolved noble gases."
Internal Threat or Hampering **(See Section 3.2.10 of this report)**	
	§ 50.73(a)(2)(x) "Any event that posed an actual threat to the safety of the nuclear power plant or significantly hampered site personnel in the performance of duties necessary for the safe operation of the nuclear power plant including fires, toxic gas releases, or radioactive releases."
Transport of a Contaminated Person Offsite **(See Section 3.2.11 of this report)**	
§ 50.72(b)(3)(xii) "Any event requiring the transport of a radioactively contaminated person to an offsite medical facility for treatment."	
News Release or Notification of Other Government Agency **(See Section 3.2.12 of this report)**	
§ 50.72(b)(2)(xi) "Any event or situation, related to the health and safety of the public or onsite personnel, or protection of the environment, for	

Table 1 Reportable Events (continued)

which a news release is planned or notification to other government agencies has been or will be made. Such an event may include an onsite fatality or inadvertent release of radioactively contaminated materials."	
Loss of Emergency Preparedness Capabilities **(See Section 3.2.13 of this report)**	
§ 50.72(b)(3)(xiii) "Any event that results in a major loss of emergency assessment capability, offsite response capability, or offsite communications capability (e.g., significant portion of control room indication, emergency notification system, or offsite notification system)."	
Single Cause that Could Have Prevented Fulfillment of the Safety Functions of **Trains or Channels in Different Systems (See Section 3.2.14 of this report)**	
	§ 50.73(a)(2)(ix)(A) "Any event or condition that as a result of a single cause could have prevented the fulfillment of a safety function for two or more trains or channels in different systems that are needed to: (1) Shut down the reactor and maintain it in a safe shutdown condition; (2) Remove residual heat; (3) Control the release of radioactive material; or (4) Mitigate the consequences of an accident." **§ 50.73(a)(2)(ix)(B)** "Events covered in paragraph (ix)(A) of this section may include cases of procedural error, equipment failure, and/or discovery of a design, analysis, fabrication, construction, and/or procedural inadequacy. However, licensees are not required to report an event pursuant to paragraph (ix)(A) of this section if the event results from: (1) A shared dependency among trains or channels that is a natural or expected consequence of the approved plant design; or (2) Normal and expected wear or degradation."

2. REPORTING AREAS WARRANTING SPECIAL MENTION

This section clarifies specific areas that are applicable to multiple reporting criteria or that historically appear to be subject to varied interpretations.

2.1 Engineering Judgment

The reportability of many events and conditions is self evident. However, the reportability of other events and conditions may not be readily apparent, and the use of engineering judgment is involved in determining reportability.

Engineering judgment may include either a documented engineering analysis or a judgment by a technically qualified individual, depending on the complexity, seriousness, and nature of the event or condition. A documented engineering analysis is not a requirement for all events or conditions, but it would be appropriate for particularly complex situations. In addition, although it is not required by the rule, it may be prudent to record in writing that a judgment was exercised by identifying the individual making the judgment, the date made, and briefly documenting the basis for this judgment. In any case, the staff considers that the use of engineering judgment implies a logical thought process that supports the judgment.

2.2 Differences in Tense between 10 CFR 50.72 and 50.73

The present tense is generally used in 10 CFR 50.72 because the event or condition generally would be ongoing at the time of reporting. The past tense is used in 10 CFR 50.73 because the event or condition is generally past when an LER is written. However, unless otherwise specified, events that occurred within 3 years of the date of discovery are reportable under 10 CFR 50.72 and 50.73 regardless of the plant mode or power level and regardless of the significance of the structure, system, or component (SSC) that initiated the event. Specific criteria in Section 3 of this report contain additional details for when tense, plant mode, power level, and significance of the SSC that initiated the event are relevant to reportability.

2.3 Reporting Multiple Events in a Single Report

More than one failure or event may be reported in a single ENS notification or LER if (1) the failures or events are related (i.e., they have the same general cause or consequences) and (2) they occurred during a single activity (e.g., a test program) over a reasonably short time (e.g., within 4 hours or 8 hours for ENS notifications, or within 60 days for LER reporting).

To the extent feasible, report failures that occurred within the first 60 days of discovery of the first failure in one LER. If appropriate, state in the LER text that a supplement to the LER will be submitted when the test program is completed. In the revised LER, include all of the failures, including those reported in the original LER (i.e., the revised LER should stand alone).

Generally, LERs are intended to address specific events and plant conditions. Thus, unrelated events or conditions should not be reported in one LER. Also, an LER revision should not be used to report subsequent failures of the same or like components that are the result of a different cause or for separate events or activities.

Unrelated failures or events should be reported as separate ENS notifications to be given unique ENS numbers by the NRC. However, multiple ENS notifications may be addressed in a single telephone call.

2.4 Deficiencies Discovered during Engineering Reviews or Inspections

As indicated in NUREG-1397, "An Assessment of Design Control Practices and Design Reconstitution Programs in the Nuclear Power Industry," issued February 1991, Section 4.3.2, the reporting requirements specified in 10 CFR 50.9, "Completeness and Accuracy of Information," 10 CFR 50.72, and 10 CFR 50.73 apply equally to discrepancies discovered during design document reconstitution programs, design-bases documentation reviews, and other similar engineering reviews. There is no basis for treating discrepancies discovered during such reviews differently from any other reportable item.

Licensees should evaluate the reportability of suspected but unsubstantiated discrepancies discovered during such a review program in the same manner as other potentially reportable items. See Section 2.5 for a discussion of reporting time limits and discovery dates.

2.5 Time Limits for Reporting

Reporting times in 10 CFR 50.72 are keyed to the occurrence of the event or condition, as described below. The present tense is generally used in 10 CFR 50.72 because the event or condition generally would be ongoing at the time of reporting. If a reportable event or condition is not on-going at the time of discovery, guidance on time limits for reporting under 10 CFR 50.73 at the end of this section may be used for determining the time of discovery for 10 CFR 50.72 reports.

In 10 CFR 50.72, the NRC requires an ENS notification within the following times:

- Paragraph 50.72(a)(3) requires ENS notification of the declaration of an Emergency Class "immediately after notification of the appropriate State or local agencies and not later than one hour after the time the licensee declares one of the Emergency Classes."

- Paragraph 50.72(b)(1) requires ENS notification for one type of event "as soon as practical and in all cases within one hour of the occurrence of any deviation from the plant's technical specifications authorized."

- Paragraph 50.72(b)(2) requires ENS notification for specific types of events and conditions "as soon as practical and in all cases, within four hours of the occurrence of any of the following."

- Paragraph 50.72(b)(3) requires ENS notification for specific types of events and conditions "as soon as practical and in all cases, within eight hours of the occurrence of any of the following."

These 10 CFR 50.72 reporting times have some flexibility because a licensee needs to ensure that reporting does not interfere with plant operation. However, that does not mean that a licensee should automatically wait until close to the time limit expiration before reporting.

In 10 CFR 50.73, the NRC requires submittal of an LER "within 60 days after the discovery" of a reportable event. Many reportable events are discovered when they occur. However, if the event is discovered at some later time, the discovery date is when the reportability clock starts under 10 CFR 50.73.

The discovery date is generally the date when the event was discovered rather than the date when an evaluation of the event is completed. For example, if a technician sees a problem, but a delay occurs before an engineer or supervisor has a chance to review the situation, the discovery date (which starts the 60-day clock) is the date that the technician sees a problem.

However, in some cases, such as discovery of an existing but previously unrecognized condition, it may be necessary to undertake an evaluation in order to determine if an event or condition is reportable. If so, the guidance provided in Regulatory Issue Summary (RIS) 2005-20, Revision 1, "Revision to NRC Inspection Manual Part 9900 Technical Guidance, 'Operability Determinations & Functionality Assessments for Resolution of Degraded or Nonconforming Conditions Adverse to Quality or Safety,'" dated April 16, 2008, which applies primarily to operability determinations, is appropriate for reportability determinations as well. This guidance indicates that the evaluation should proceed on a time scale commensurate with the safety significance of the issue and that, whenever reasonable expectation no longer exists that the equipment in question is operable, or significant doubts begin to arise, appropriate actions, including reporting, should be taken.

2.6 Events Discussed with the NRC Staff

On occasion, some licensee personnel have erroneously believed that if a reportable event or condition had been discussed with the resident inspector or other NRC staff, there was no need to report under 10 CFR 50.72 and 50.73 because the NRC was aware of the situation. Some licensee personnel have also expressed a similar misunderstanding for cases in which the NRC staff identified a reportable event or condition to the licensee via inspection or assessment activities. Such conditions do not satisfy 10 CFR 50.72 and 50.73. The regulations in 10 CFR 50.72 and 50.73 specifically require a telephone notification via the ENS and/or submittal of a written LER for an event or condition that meets the criteria stated in those rules.

2.7 Voluntary Reporting

Information that does not meet the reporting criteria of 10 CFR 50.72 and 50.73 may be reportable under other requirements, such as 10 CFR 50.9, 20.2202, 20.2203, 50.36, 72.74, 72.216, 73.71, and 10 CFR Part 21, "Reporting of Defects and Noncompliance." In particular, 10 CFR 50.9(b) states that "Each applicant or licensee shall notify the Commission of information identified by the applicant or licensee as having for the regulated activity a significant implication for public health and safety or common defense and security." This applies to information that is not already required by other reporting or updating requirements. Notification must be made to the Administrator of the appropriate regional office within 2 working days of identifying the information. Reporting under 10 CFR 50.9, "Completeness and Accuracy of Information," is required, not voluntary.[1]

---

[1]   As indicated in the Statement of Considerations for 10 CFR 50.9, "A licensee cannot evade the rule by never 'finding' information to be significant. The fact that a licensee considers information to be significant can be established, for example, by the actions taken by the licensee to evaluate that information" (59 FR 49362; December 31, 1987).

Voluntary reporting, as discussed in the following paragraphs, pertains to information of lesser significance than that described in 10 CFR 50.9(b).

Licensees are permitted and encouraged to report any event or condition that does not meet the criteria for required reporting, if the licensee believes that the event or condition might be of safety significance or of generic interest or concern. Reporting requirements aside, assurance of safe operation of all plants depends on accurate and complete reporting by each licensee of all events having potential safety significance. Instructions for voluntary ENS notifications and LERs are discussed in Sections 4.2.2 and 5.1.4 of this report.

The NRC staff encourages voluntary LERs rather than information letters for voluntary reporting. The LER format is preferable because it provides the information needed to support NRC review of the event and facilitates administrative processing, including data entry.

2.8 Retraction or Cancellation of Event Reports

An ENS notification may be retracted via a followup telephone call, as discussed further in Section 4.2.3 of this report. A retracted ENS report is retained in the ENS database, along with the retraction.

An LER may be canceled by letter as discussed further in Section 5.1.2 of this report. Canceled LERs are deleted from the LER database.

Sound, logical bases for the withdrawal should be communicated with the retraction or cancellation. (Example 3 in Section 3.2.4 of this report illustrates a case in which there were sound reasons for a retraction. The last event under Example 1 in Section 3.2.6 illustrates a case in which the reasons for retraction were not adequate.)

For reports that were submitted as a result of an operability determination, the retraction or cancellation should discuss why the operability determination was revised, as well as its impact on the associated reporting criteria (e.g., system operability was never lost, the limiting condition for operation (LCO) was always met, completion times were never exceeded, etc). RIS 2005-20, Revision 1 provides guidance on operability determinations.

3. SPECIFIC REPORTING GUIDELINES

3.1 General Requirements

3.1.1 Immediate Notifications

§ 50.72(a) "General requirements.[1]	§ 50.73
(1) Each nuclear power reactor licensee licensed under § 50.21(b) or § 50.22 of this part shall notify the NRC Operations Center via the Emergency Notification System of: (i) The declaration of any of the Emergency Classes specified in the licensee's approved Emergency Plan;[2] or (ii) Those non-emergency events specified in paragraph (b) of this section that occurred within three years of the date of discovery. (2) If the Emergency Notification System is inoperative, the licensee shall make the required notifications via commercial telephone service, other dedicated telephone system, or any other method which will ensure that a report is made as soon as practical to the NRC Operations Center.[3] (3) The licensee shall notify the NRC immediately after notification of the appropriate State or local agencies and not later than one hour after the time the licensee declares one of the Emergency Classes. (4) The licensee shall activate the Emergency Response Data System (ERDS)[4] as soon as possible but not later than one hour after declaring an Emergency Class of alert, site area emergency, or general emergency. The ERDS may also be activated by the licensee during emergency drills or exercises if the licensee's computer system has the capability to transmit the exercise data. (5) When making a report under paragraph (a)(1) of this section, the licensee shall identify: (i) The Emergency Class declared; or (ii) Paragraph (b)(1), 'One-hour reports,' paragraph (b)(2) 'Four hour reports,' or paragraph (b)(3), 'Eight	There is no requirement in 10 CFR 50.73 to report the declaration of an emergency class. However, an event or condition that leads to declaration of an emergency class may meet one or more of the specific reporting requirements that are in 10 CFR 50.73. There is usually a parallel reporting requirement in 10 CFR 50.73 that captures a nonemergency event that is reportable under 10 CFR 50.72. Exceptions are a press release, notification to another government agency, transport of a contaminated person offsite, and loss of emergency preparedness capability.

hour reports,' as the paragraph of this section requiring notification of the non-emergency event."	
¹ Other requirements for immediate notification of the NRC by licensed operating nuclear power reactors are contained elsewhere in this chapter, in particular, §§ 20.1906, 20.2202, 50.36, 72.216 and 73.71. ² These Emergency Classes are addressed in Appendix E of this part. ³ Commercial telephone number of the NRC Operations Center is (301) 816-5100. ⁴ Requirements for ERDS are addressed in Appendix E, Section VI.	

Discussion

Appendix E, "Emergency Planning and Preparedness for Production and Utilization Facilities," to 10 CFR Part 50, "Domestic Licensing of Production and Utilization Facilities," Section IV.C, "Activation of Emergency Organization," establishes four emergency classes for nuclear power plants: (1) notification of unusual event, (2) alert, (3) site area emergency, and (4) general emergency. NUREG-0654/FEMA-REP-1, Revision 1, "Criteria for Preparation and Evaluation of Radiological Emergency Response Plans and Preparedness in Support of Nuclear Power Plants," issued November 1980; Nuclear Management and Resources Council/National Environmental Studies Project (NUMARC/NESP)-007, Revision 2, "Methodology for Development of Emergency Action Levels," issued January 1992; and NEI 99-01, "Methodology for Development of Emergency Action Levels," provide the basis for these emergency classes and numerous examples of the events and conditions typical of each emergency class. Licensees use this guidance in preparing their emergency plans. Use of these four emergency classification terms in the ENS notification helps the NRC recognize the significance of an emergency. Timeframes specified for notification in 10 CFR 50.72(a) use the words "immediately" and "not later than one hour" to ensure that the Commission can fulfill its responsibilities during and following the most serious events.

Occasionally, a licensee discovers that a condition existed that met the emergency plan criteria but no emergency was declared and the basis for the emergency class no longer exists at the time of this discovery. This may be due to a rapidly concluded event or an oversight in the emergency classification made during the event, or it may be determined during a post event review. Frequently, in cases of this nature, which were discovered after the fact when the plant conditions that would have initiated the classification and notifications are no longer present, licensees have declared the emergency class, immediately terminated the emergency class, and then made the appropriate notifications. However, the NRC staff does not consider actual declaration of the emergency class to be necessary in these circumstances. If the licensee does not declare an emergency under these circumstances, an ENS notification (or an ENS update if the event was previously reported but misclassified) within 1 hour of the discovery of the undeclared (or misclassified) event provides an acceptable alternative.[2] Nonetheless, if the licensee does declare an emergency, then all notifications required by 10 CFR 50.47(b)(5), 10 CFR 50.72, and Appendix E to 10 CFR Part 50, Section IV.D.3 are to be made.

3.1.2 Licensee Event Reports

§ 50.72	§ 50.73(a)(1)
There is no comparable passage in 10 CFR 50.72.	"The holder of an operating license for a nuclear power plant (licensee) shall submit a Licensee Event Report (LER) for any event of the type described in this paragraph within 60 days after the discovery of the event. In the case of an invalid actuation reported under § 50.73(a)(2)(iv), other than actuation of the reactor protection system (RPS) when the reactor is critical, the licensee may, at its option, provide a telephone notification to the NRC Operations Center within 60 days after discovery of the event instead of submitting a written LER. Unless otherwise specified in this section, the licensee shall report an event if it occurred within three years of the date of discovery regardless of the plant mode or power level, and regardless of the significance of the structure, system, or component that initiated the event."

Discussion

Unless otherwise specified, this part of the rule requires reporting of an event if it occurred within 3 years prior to discovery, regardless of the plant mode or power level and regardless of the significance of the SSC that initiated the event. In the case of an invalid actuation reported under 10 CFR 50.73(a)(2)(iv), the licensee may, at its option, provide a telephone notification to the NRC Operations Center within 60 days after discovery of the event instead of submitting a written LER.

3.2 Specific Reporting Criteria

3.2.1 Plant Shutdown Required by Technical Specifications

§ 50.72(b)(2)(i)	§ 50.73(a)(2)(i)(A)
"The initiation of any nuclear plant shutdown required by the plant's Technical Specifications."	"The completion of any nuclear plant shutdown required by the plant's Technical Specifications; or"

---

[2] The licensee should inform State and local offsite response organizations of such events in accordance with the arrangements made between the licensee and offsite organizations.

If not reported under 10 CFR 50.72(a) or (b)(1), an ENS notification is required. The initiation of any nuclear plant shutdown required by the plant's Technical Specifications (TS) should be apparent at the time of occurrence. Therefore, if all events are reported properly, it is expected that all reports under 10 CFR 50.72 are as a result of an on-going condition. If the shutdown is completed, an LER is required.

Discussion

The 10 CFR 50.72 reporting requirement is intended to capture those events for which TS require the initiation of reactor shutdown to provide the NRC with early warning of safety-significant conditions serious enough to warrant that the plant be shut down. For 10 CFR 50.72 reporting purposes, the phrase "initiation of any nuclear plant shutdown" includes action to start reducing reactor power; i.e., adding negative reactivity to achieve a nuclear plant shutdown required by TS. This includes initiation of any shutdown due to expected inability to restore equipment prior to exceeding the LCO action time. As a practical matter, in order to meet the time limits for reporting under 10 CFR 50.72, the reporting decision should sometimes be based on such expectations. (See Example 4.)

The "initiation of any nuclear plant shutdown" does not include mode changes required by TS if they are initiated after the plant is already in a shutdown condition.

A reduction in power for some other purpose, not constituting initiation of a shutdown required by TS, is not reportable under this criterion.

For 10 CFR 50.73 reporting purposes, the phrase "completion of any nuclear plant shutdown" is defined as the point in time during a TS-required shutdown when the plant enters the first shutdown condition required by an LCO (e.g., hot standby (Mode 3) for PWRs with the Standard Technical Specifications (STS)). For example, if at 0200 hours a plant enters an LCO action statement that states, "restore the inoperable channel to operable status within 12 hours or be in at least Hot Standby within the next 6 hours," the plant must be shut down (i.e., at least in hot standby) by 2000 hours. An LER is required if the inoperable channel is not returned to operable status by 2000 hours and the plant enters hot standby.

An LER is not required if a failure was or could have been corrected before a plant has completed shutdown (as discussed above) and no other criteria in 10 CFR 50.73 apply.

Examples

(1)   Initiation of a TS-Required Plant Shutdown

While operating at 100-percent power, one of the battery chargers, which feeds a 125-volt direct current vital bus, failed during a surveillance test. The battery charger was declared inoperable, placing the plant in a 2-hour LCO to return the battery charger to an operable status or commence a TS-required plant shutdown. Licensee personnel started reducing reactor power to achieve a nuclear plant shutdown required by a TS when they were unable to complete repairs to the inoperable battery charger in the 2 hours allowed. The cause of the battery charger failure was subsequently identified and repaired. Upon completion of surveillance testing, the battery charger was returned to service and the TS-required plant shutdown was stopped at 96-percent power.

The licensee made an ENS notification because of the initiation of a TS-required plant shutdown. An LER was not required under this criterion because the failed battery charger was corrected before the plant completed shutdown.

(2) <u>Initiation and Completion of a TS-Required Plant Shutdown</u>

During startup of a PWR plant with reactor power in the intermediate range, two of the four reactor coolant pumps tripped when the station power transformer supplying power de-energized. With less than four reactor coolant pumps operating, the plant entered a 1-hour LCO to be in hot standby. Control rods were manually inserted to place the plant in a shutdown condition.

The licensee made an ENS notification because of the initiation of a TS-required plant shutdown. An LER was required because of the completion of the TS-required plant shutdown.

(3) <u>Failure that Was or Could Have Been Corrected Before Shutdown Was Required</u>

• Question: What about the situation where you have 7 days to fix a component or be shut down, but the plant must be shut down to fix the component? Assume the plant shuts down, the component is fixed, and the plant returns to power before the end of the 7-day period. Is that situation reportable?

Answer: No. If the shutdown was not required by the TS, it need not be reported. However, other criteria in 50.73 may apply and may require that the event be reported.

• Question: Suppose that there are 7 days to fix a problem, and it is likely that the problem can be fixed during this time period. However, the plant management elects to shut down and fix this problem and other problems. Is an LER required?

Answer: No. Some judgment is required. The shutdown is reportable, however, if the situation could not have been corrected before the plant was required to be shut down.

(4) <u>Initiation of Plant Shutdown in Anticipation of an LCO-Required Shutdown</u>

The plant lost one of two sources of offsite power due to overheating in the main transformer. The TS allow 72 hours to restore the source or initiate a shutdown and be in hot standby within the next 6 hours and cold shutdown within the following 30 hours. The licensee estimated that the transformer problem could not be corrected within the LCO action time. Therefore, the decision was made to start a shutdown soon after the transformer problem was discovered.

The shutdown was uneventful and was completed, with the plant in hot standby, before the expiration of the LCO action time. After the plant reached hot standby, further evaluation indicated that the transformer problem could not be corrected before the requirement to place the plant in cold shutdown. Based on this time estimate, it was decided to place the unit in cold shutdown.

The event is reportable under 10 CFR 50.72(b)(2)(i) as the initiation of a plant shutdown required by TS because, at the time the shutdown was initiated and the time the report was due, it was not expected that the equipment would be restored to operable status within the required time. This is based on the fact that the reporting requirement is intended to capture those events for which TS require the initiation of a reactor shutdown.

The event is reportable under 10 CFR 50.73(a)(2)(i)(A) because the plant shutdown was completed when the plant reached hot standby (Mode 3). Had the transformer been repaired and the shutdown process terminated before the plant reached Mode 3, the event would not be reportable under 10 CFR 50.73(a)(2)(i)(A).

3.2.2 Operation or Condition Prohibited by Technical Specifications

§ 50.72	§ 50.73(a)(2)(i)(B)
There is no corresponding requirement in 10 CFR 50.72.	"Any operation or condition which was prohibited by the plant's Technical Specifications except when: (1) The Technical Specification is administrative in nature; (2) The event consisted solely of a case of a late surveillance test where the oversight was corrected, the test was performed, and the equipment was found to be capable of performing its specified safety functions; or (3) The Technical Specification was revised prior to discovery of the event such that the operation or condition was no longer prohibited at the time of discovery of the event."

An LER is required for any operation or condition that was prohibited by the plant's TS, subject to the exceptions stated in the rule. The NRC expects licensees to include violations of the TS in their corrective action programs, which are subject to NRC audit.

Discussion[3]

Safety Limits and Limiting Safety System Settings

The regulation in 10 CFR 50.36(c)(1) outlines the reporting requirements in TS for events in which safety limits or limiting safety system settings are exceeded. It indicates that such reports are to be made as required by 10 CFR 50.72 and 50.73. There would not be a 3-year limitation in this case because, in addition to the requirements of 10 CFR 50.72 and 50.73, specific reporting requirements are stated in 10 CFR 50.36(c)(1), and perhaps in the plant's TS.

---

[3] This criterion does not address violations of license conditions that are contained in documents other than the TS. Such violations are reportable as specified in the plant's license or other applicable documents.

Limiting Conditions for Operation

The regulation in 10 CFR 50.36(c)(2) outlines LCOs in TS. Certain TS contain LCO statements that include action statements (required actions and their associated completion time in the STS) to provide constraints on the length of time components or systems may remain inoperable or out of service before the plant must shut down or other compensatory measures must be taken. Such time constraints are based on the safety significance of the component or system being removed from service.

An LER is required if a condition existed for a time longer than permitted by the TS (i.e., greater than the total allowed restoration and shutdown outage time (or completion time in the STS)), even if the condition was not discovered until after the allowable time had elapsed and the condition was rectified immediately upon discovery. This guidance is consistent with that previously given. (For the purpose of this discussion, it is assumed that there was firm evidence that a condition prohibited by TS existed before discovery, for a time longer than permitted by TS.)

Technical Specification Surveillance Testing

The regulation in 10 CFR 50.36(c)(3) outlines surveillance requirements in TS that assure (1) the necessary quality of systems and components, (2) operation within safety limits, and (3) meeting the LCOs.

Generally, an operation or condition prohibited by the TS existed and is reportable if surveillance testing indicates that equipment (e.g., one train of a multiple-train system) was not capable of performing its specified safety functions (and thus was inoperable) for a period of time longer than allowed by TS (i.e., the LCO-allowed outage time, or the completion time for restoration of equipment in the STS). Reporting is not required if an event consists solely of a case of a late surveillance test in which the oversight is corrected, the test is performed, and the equipment is found to be capable of performing its specified safety functions.

For the purpose of evaluating the reportability of a discrepancy found during surveillance testing that is required by the TS, licensees should do the following:

(1)   For testing that is conducted within the required time (i.e., the surveillance interval plus any extension allowed by STS Surveillance Requirement (SR) 3.0.2 or its equivalent), it should be assumed that the discrepancy occurred at the time of its discovery unless there is firm evidence, based on a review of relevant information such as the equipment history and the cause of failure, to indicate that the discrepancy existed previously.

(2)   For testing that is conducted later than the required time, it should be assumed that the discrepancy occurred at the time the testing was required unless there is firm evidence to indicate that it occurred at a different time.

The purpose of this approach is twofold. First, it rules out reporting of routine occurrences (i.e., occurrences for which a timely surveillance test is performed, the results fall outside of acceptable limits, and the condition is corrected) unless there is firm evidence that equipment was incapable of performing its specified safety function longer than allowed. On the other hand, if the surveillance test is performed substantially late, and the equipment is not capable of performing its specified safety function, the occurrence is not routine. In this case, the event is reportable unless there is firm evidence that the duration of the discrepancy was within allowed limits.

For cases in which it is discovered that a surveillance test was not performed within its specified frequency or interval, some plants have TS that allow a delay in declaring an LCO or that TS requirements were not met (i.e., STS SR 3.0.3 or its equivalent). This allows time to perform the test before making such a declaration and taking other required actions. However, an LER would still be required if the test indicates that equipment (e.g., one train of a multiple-train system) was not capable of performing its specified safety functions (and thus was inoperable) for a period of time longer than allowed by TS. The allowed delay in declaring the LCO not met does not change the fact that the condition existed longer than allowed by TS.

Tests Required by Section XI of the American Society of Mechanical Engineers Code

In 10 CFR 50.55a(g) and 50.55a(f), the NRC requires the implementation of inservice inspection and inservice testing programs in accordance with the applicable edition of the American Society of Mechanical Engineers (ASME) Code for those pumps and valves whose function is required for safety. The STS contain these testing requirements.

As with surveillance testing, an operation or condition prohibited by the TS existed and is reportable if the testing indicates that equipment (e.g., one train of a multiple-train system required to be operable by the TS) was not capable of performing its specified safety functions (and thus was inoperable) for a period of time longer than allowed by TS. Accordingly, similar assumptions and standards should be used. For example, if a timely test indicates that equipment is not capable of performing its specified safety function, it should be assumed that the discrepancy occurred at the time of the test unless there is firm evidence to indicate that it existed previously.

Design and Analysis Defects and Deviations

A design or analysis defect or deviation is reportable under this criterion if, as a result, equipment (e.g., one train of a multiple-train system) was not capable of performing its specified safety functions (and thus was inoperable) for a period of time longer than allowed by TS. Because design and analysis conditions are long lasting, the essential question in this case is whether the equipment was capable of performing its specified safety functions.

Administrative Requirements

Section 5 of the STS, or its equivalent, has a number of administrative requirements, such as organizational structure, the required number of personnel on shift, the maximum hours of work permitted during a specific interval of time, and the requirement to have, maintain, and implement certain specified procedures. Violation of a TS that is administrative in nature is not reportable.

For example, a change in the plant's organizational structure that has not yet been approved as a TS change would not be reportable. An administrative procedure violation, or failure to implement a procedure, such as failure to lock a high-radiation-area door, is generally not reportable under this criterion. Radiological conditions and events that are reportable are defined in 10 CFR 20.2202, "Notification of Incidents," and 10 CFR 20.2203, "Reports of Exposures, Radiation Levels, and Concentrations of Radioactive Material Exceeding the Constraints or Limits." Redundant reporting is not required.

<u>Entry into STS 3.0.3</u>

STS LCO 3.0.3, or its equivalent, establishes requirements for actions when (1) an LCO is not met and the associated actions are not met, (2) an associated action is not provided, or (3) as directed by the associated actions themselves.

Entry into LCO 3.0.3 or its equivalent is not necessarily reportable under this criterion. However, it should be considered reportable under this criterion if any of the shutdown times listed in LCO 3.0.3 (e.g., Modes 3, 4, or 5 for the Westinghouse STS) were exceeded, even if the condition was not discovered until after the allowable time had elapsed and the condition was rectified immediately upon discovery.

For a given LCO condition, if shutdown required actions and completion times are listed (e.g., be in hot shutdown in X hours and cold shutdown in Y hours), shutdown times associated with LCO 3.0.3 should not be considered (i.e., only consider LCO action table shutdown completion time added to restorative completion time when determining if "Operations or Conditions Prohibited by Technical Specifications" existed). If entry into LCO 3.0.3 is explicitly listed as a required action for a given condition, or for cases in which the condition is not listed in the action table, shutdown times associated with LCO 3.0.3 may be added to any associated restorative completion times found in the action table when determining if "Operations or Conditions Prohibited by Technical Specifications" existed.

The discussion contained in this section only pertains to "Operations or Conditions Prohibited by Technical Specifications." Entry into LCO 3.0.3 may still result in other reportable conditions under 10 CFR 50.72 and 50.73.

<u>Revised Technical Specifications</u>

An LER is not required for discovery of an operation or condition that occurred in the past and was prohibited at the time it occurred if, before the time of discovery, the TS were revised such that the operation or condition is no longer prohibited. Such an event would have little or no significance because the operation or condition would have been determined to be acceptable and allowed under the current TS.

Examples

(1)   <u>Limiting Condition for Operation Exceeded</u>

In conducting a timely 30-day surveillance test, a licensee found a standby component with a 7-day LCO-allowed outage time and an associated 8-hour shutdown action statement to be inoperable. (This is equivalent to a 7-day restoration completion time and an 8-hour action completion time in the STS.) Subsequent review indicated that the component was assembled improperly during maintenance conducted 30 days previously and the post maintenance test was not adequate to identify the error. Thus, there was firm evidence that the standby component had been inoperable for the entire 30 days.

An LER was required because the condition existed longer than allowed by the TS (the 7-day LCO-allowed outage time and the shutdown action statement time of 8 hours). Had the inoperability been identified and corrected within the required time, the event would not be reportable.

(2)   Late Surveillance Tests

A licensee, with the plant in Mode 5 following a 10-month refueling outage, determined that certain monthly TS surveillance tests, which were required to be performed regardless of plant mode, had not been performed as required during the outage. The STS SR 3.0.2 extension was also exceeded. The surveillance tests were immediately performed.

No LER would be required if the tests showed the equipment was still capable of performing its specified safety functions. On the other hand, if the tests showed the equipment was not capable of performing its specified safety functions (and thus was inoperable) in excess of the allowed time, the event would be reportable.

(3)   Multiple Test Failures

An example of multiple test failures involves the sequential testing of safety valves. Sometimes multiple valves are found to lift with setpoints outside of TS limits.

As discussed above, discrepancies found in TS surveillance tests should be assumed to occur at the time of the test unless there is firm evidence, based on a review of relevant information (e.g., the equipment history and the cause of failure), to indicate that the discrepancy occurred earlier. However, the existence of similar discrepancies in multiple valves is an indication that the discrepancies may well have arisen over a period of time and that the failure mode should be evaluated to make this determination. If so, the condition existed during plant operation and the event is reportable under 10 CFR 50.73(a)(2)(i)(B) ("Any operation or condition prohibited by the plant's Technical Specifications").

If the discrepancies are large enough that multiple valves are inoperable, the event may also be reportable under the following criterion in 10 CFR 50.73(a)(2)(vii): "Any event where a single cause or condition caused at least one independent train or channel to become inoperable in multiple systems or two independent trains or channels to become inoperable in a single system."

(4)   Seismic Restraints

Assume it is found that an exciter panel for one EDG had lacked appropriate seismic restraints since the plant was constructed because of a design, analysis, or construction inadequacy. Upon evaluation, the EDG is determined to be inoperable because it is not capable of performing its specified safety functions during and after safe-shutdown earthquake.

An LER would be required because the EDG was inoperable for a period of time longer than allowed by TS.

(5)   Vulnerability to Loss of Offsite Power

Assume that during a design review it is found that a loss of offsite power could cause a loss of instrument air and, as a result, auxiliary feedwater (AFW) flow control valves could fail open. Therefore, for low steam generator pressure such as could occur for certain main steamline breaks, high AFW flow rates could result in tripping the motor-driven AFW pumps on thermal overload. Therefore, the motor-driven AFW pumps are determined to be inoperable. The single turbine-driven AFW pump is not affected.

An LER would be required because the motor-driven portion of AFW was inoperable for a period of time longer than allowed by the TS.

(6)   Entry into STS 3.0.3

The following two examples illustrate the "Entry into STS 3.0.3" discussions. Both examples use STS 3.5.1, "Accumulators" found in Westinghouse STS, Revision 4, (Agencywide Documents Access and Management System (ADAMS) Accession No. ML12100A222), and assumes historic and current operations in Mode 1.

(a) One accumulator inoperable due to boron concentration not within limits

When evaluating if an Operation or Condition Prohibited by TS occurred, the timeframe to consider is the total time allowed by Condition A (restorative Completion Time) added to Condition C (shutdown Completion Time). Condition A, "One accumulator inoperable due to boron concentration not within limits," has a Completion Time of 72 hours, while Condition C, "Required Action and associated Completion Time of Condition A or B not met," has a Completion Time of 6 hours to be in Mode 3. If it is determined that one accumulator was inoperable due to boron concentration not within limits for greater than 78 hours (72 hour restorative Completion Time added to 6 hour shutdown Completion Time), then an Operation or Condition Prohibited by TS existed. The condition would be reportable even if the condition was not discovered until after the allowable time had elapsed and the condition was rectified immediately upon discovery. Since the Action table contains shutdown required actions and completion times for the given condition, times associated with LCO 3.0.3 are not considered.

(b) Two or more accumulators inoperable

When evaluating if an Operation or Condition Prohibited by TS occurred, the timeframe to consider is the total time allowed by Condition D (restorative Completion Time) added to the shutdown time associated with LCO 3.0.3. Condition D, "Two or more Accumulators inoperable," has a Completion Time of Immediately and LCO 3.0.3 requires that the unit be in Mode 3 within 7 hours. If it is determined that two or more accumulators were inoperable for greater than 7 hours (Immediate restorative Completion Time added to 7 hour shutdown time), then an Operation or Condition Prohibited by TS existed. The condition would be reportable even if the condition was not discovered until after the allowable time had elapsed and the condition was rectified immediately upon discovery.

(7) Laboratory Testing of Charcoal Adsorbers

The following example illustrates a scenario in which the results of testing required by the TS are available at a later time. The example uses STS 3.7.10, "Control Room Emergency Filtration System (CREFS)" found in Westinghouse STS, Revision 4, (ADAMS Accession No. ML12100A222).

While operating in Mode 1, SR 3.7.10.2 is performed on two CREFS filter trains. SR 3.7.10.2 requires CREFS filter testing in accordance with the Ventilation Filter Testing Program (VFTP) found in TS 5.5.11. Part c of the VFTP requires laboratory testing of charcoal adsorber samples. On Day 1, the licensee takes samples and sends it out to a laboratory for analysis. The plant continues to operate in Mode 1. On Day 20, the licensee gets back the results that indicate that the methyl iodide penetration is greater than the value specified in TS 5.5.11.c for one of the filter trains. The licensee then declares the effected CREFS train inoperable.

As stated in the Discussion Section above, an Operation or Condition Prohibited by TS can exist even if the condition was not discovered until after the allowable time had elapsed and the condition was rectified immediately upon discovery. When evaluating if an Operation or Condition Prohibited by TS occurred, the timeframe to consider is the total time allowed by Condition A (restorative Completion Time) added to Condition C (shutdown Completion Time). Condition A, "One CREFS train inoperable for reasons other than Condition B {inoperable boundary}," has a Completion Time of 7 days and Condition C, "Required Action and associated Completion Time of Condition A or B not met..." has a Completion Time of 6 hours to be in Mode 3.

Since the inoperable condition on the effected CREFS train existed for at least 20 days, an Operation or Condition Prohibited by TS existed (i.e., the inoperable condition exceeded the combined 7 day restorative Completion Time added to the 6 hour shutdown Completion Time),

The discussions found in this example would still be relevant even if plant specific TS provide additional restrictions on when the test sample must be received back from the laboratory (i.e. verify results within 31 days after sample removal, etc).

3.2.3 Deviation from Technical Specifications under 10 CFR 50.54(x)

§ 50.72(b)(1)	§ 50.73(a)(2)(i)(C)
"any deviation from the plant's Technical Specifications authorized pursuant to § 50.54(x) of this part."	"Any deviation from the plant's Technical Specifications authorized pursuant to § 50.54(x) of this part."

An LER is required for a deviation authorized under 10 CFR 50.54(x). If not reported under 10 CFR 50.72(a), an ENS notification is also required. Any deviation from the plant's TS authorized pursuant to § 50.54(x) should be apparent at the time of occurrence. Therefore, if all events are reported properly, it is expected that all reports under 10 CFR 50.72 are as a result of an on-going condition.

Discussion

In 10 CFR 50.54(x), the NRC generally permits licensees to take reasonable action in an emergency even though the action departs from the plant TS if (1) the action is immediately needed to protect the public health and safety, including plant personnel, and (2) no action consistent with the TS is immediately apparent that can provide adequate or equivalent protection. TS deviations authorized under 10 CFR 50.54(x) are reportable under this criterion.

3.2.4 Degraded or Unanalyzed Condition

§ 50.72(b)(3)(ii)	§ 50.73(a)(2)(ii)
"Any event or condition that results in:	Any event or condition that resulted in:
(A) The condition of the nuclear power plant, including its principal safety barriers, being seriously degraded; or	(A) The condition of the nuclear power plant, including its principal safety barriers, being seriously degraded;
(B) The nuclear power plant being in an unanalyzed condition that significantly degrades plant safety."	(B) The nuclear power plant being in an unanalyzed condition that significantly degraded plant safety."

An LER is required for a seriously degraded principal safety barrier or an unanalyzed condition that significantly degrades plant safety. If not reported under 10 CFR 50.72(a), (b)(1), or (b)(2), an ENS notification is required under 10 CFR 50.72(b)(3) (an 8-hour report). On occasion, a "Degraded or Unanalyzed Condition" is discovered to have occurred in the past, but is not on-going at the time of discovery. ENS notifications and LERs are required if a Degraded or Unanalyzed Condition occurred within 3 years of the date of discovery, even if the event is not on-going at the time of discovery.

Discussion

(A) Nuclear Power Plant, Including Its Principal Safety Barriers, Being Seriously Degraded

This criterion applies to material (e.g., metallurgical or chemical) problems that cause abnormal degradation of or stress upon the principal safety barriers (i.e., the fuel cladding, reactor coolant system (RCS) pressure boundary, or the containment). Abnormal degradation of a barrier may be indicated by the necessity of taking corrective action to restore the barrier's capability, as is the case in some of the examples discussed below. Abnormal stress upon a barrier may result from an unplanned transient, as is the case in one of the examples discussed below. The following are examples of reportable events and conditions:

(1)   Fuel cladding failures in the reactor, or in the storage pool, that exceed expected values, or that are unique or widespread, or that are caused by unexpected factors.

(2)   Welding or material defects in the primary coolant system that cannot be found acceptable under ASME Section XI, IWB-3600, "Analytical Evaluation of Flaws," or ASME Section XI, Table IWB-3410-1, "Acceptance Standards."

(3)   Serious steam generator tube degradation. A licensee's plant-specific TS contain performance criteria for steam generator tube integrity, which includes structural integrity, accident induced leakage, and operational leakage. Steam generator tube degradation is considered serious only if either the steam generator structural integrity or accident-induced leakage performance criteria are not met.

In addition, one or more steam generator tubes satisfying the tube repair criteria and not plugged or repaired in accordance with the steam generator program is not considered to be serious steam generator tube degradation and therefore is not reportable as a "Degraded or Unanalyzed Condition," as long as the structural integrity and accident-induced leakage performance criteria are both met.

(4)   Low temperature over pressure transients in which the pressure-temperature relationship violates pressure-temperature limits derived from Appendix G, "Fracture Toughness Requirements," to 10 CFR Part 50 (e.g., TS pressure-temperature curves).

(5)   Loss of containment function or integrity, including containment leak rate tests in which the total containment as-found, minimum-pathway leak rate exceeds limits in the facility's TS.[4]

(B) Unanalyzed Condition that Significantly Affects Plant Safety

The 1983 Statements of Consideration for 10 CFR 50.72 and 50.73 indicated the following with regard to an unanalyzed condition that significantly compromises plant safety:

> The Commission recognizes that the licensee may use engineering judgment and experience to determine whether an unanalyzed condition existed. It is not intended that this paragraph apply to minor variations in individual parameters, or to problems concerning single pieces of equipment. For example, at any time, one or more safety-related components may be out of service due to testing, maintenance, or a fault that has not yet been repaired. Any trivial single failure or minor error in performing surveillance tests could produce a situation in which two or more often unrelated, safety-grade components are out-of-service. Technically, this is an unanalyzed condition. However, these events should be reported only if they involve functionally related components or if they significantly compromise plant safety.[5]

When licensees are applying engineering judgment, and there is a doubt regarding whether or not to report, the Commission's policy is that licensees should make the report.[6]

For example, small voids in systems designed to remove heat from the reactor core that have been previously shown through analysis not to be safety significant need not be reported. However, the accumulation of voids that could inhibit the ability to adequately remove heat from the reactor core, particularly under natural circulation conditions, would constitute an unanalyzed condition and would be reportable.[7]

---

[4]   The LCO typically employs *La*, which is defined in Appendix J, "Primary Reactor Containment Leakage Testing for Water-Cooled Power Reactors," to 10 CFR Part 50 as the maximum allowable containment leak rate at pressure *Pa*, the calculated peak containment internal pressure related to the design-basis accident. "Minimum-pathway leak rate" means the minimum leak rate that can be attributed to a penetration leakage path; for example, the smaller of either the inboard or outboard valve's individual leak rates.

[5]   48 FR 39042, August 29, 1983, and 48 FR 33856, July 26, 1983.

[6]   48 FR 39042, August 29, 1983.

[7]   48 FR 39042, August 29, 1983, and 48 FR 33856, July 26, 1983.

In addition, voiding in instrument lines that results in an erroneous indication causing the operator to misunderstand the true condition of the plant is also an unanalyzed condition and should be reported.[8]

The level of significance of these cases generally corresponds to the inability to perform a required safety function. For instance, accumulation of voids that could inhibit the ability to adequately remove heat from the reactor core, particularly under natural circulation conditions, has an effect similar to a condition that could prevent the fulfillment of the safety function of the AFW system.

Beyond the examples given in 1983, an example of an event reportable as an unanalyzed condition that significantly degraded plant safety would be the discovery that a system required to meet the single failure criterion does not do so.

In another example, if fire barriers are found to be missing, such that the required degree of separation for redundant safe shutdown trains is lacking, the event would be reportable as an unanalyzed condition that significantly degraded plant safety. On the other hand, if a fire wrap, to which the licensee has committed, is missing from a safe shutdown train but another safe shutdown train is available in a different fire area, protected such that the required separation for safe shutdown trains is still provided, the event would not be reportable.

Examples

(1) <u>Failures of Reactor Fuel Rod Cladding Identified during Testing of Fuel Assemblies</u>

Radiochemistry data for a particular PWR indicated that a number of fuel rods had failed during the first few months of operation. Projections ranged from 6 to 12 failed rods. The end-of-cycle RCS iodine-131 activity averaged 0.025 microcuries per milliliter. Following the end-of-cycle shutdown, iodine-131 spiked to 11.45 microcuries per milliliter. The cause was a significant number of failed fuel rods. Inspections revealed that 136 of the total 157 fuel assemblies contained failed fuel (approximately 300 fuel rods had through-wall penetrations), far exceeding the anticipated number of failures. The defects were generally pinhole sized. The fuel cladding failures were caused by long-term fretting from debris that became lodged between the lower fuel assembly nozzle and the first spacer grid, resulting in penetration of the stainless-steel fuel cladding. The source of the debris was apparently a machining byproduct from the thermal shield support system repairs during the previous refueling outage.

The event is reportable because the cladding failures exceed expected values and are unique or widespread.

(2) <u>Reactor Coolant System Pressure Boundary Degradation Due to Corrosion of a Control Rod Drive Mechanism Flange</u>

While the plant was in hot shutdown, a total of six control rod drive mechanism (CRDM) reactor vessel nozzle flanges were identified as leaking. Subsequently, one of the flanges was found to be eroded and pitted. While removing the nut ring from beneath the flange, it was discovered that approximately 50 percent of one of the nut ring halves

---

[8]   48 FR 39042, August 29, 1983, and 48 FR 33856, July 26, 1983.

had corroded away and that two of the four bolt holes in the corroded nut ring half were degraded to the point that there was no bolt–thread engagement.

An inspection of the flanges and spiral wound gaskets, which were removed from between the flanges, revealed that the cause of the leaks was the gradual deterioration of the gaskets from age. A replacement CRDM was installed and the gaskets on all six CRDMs were replaced with new-design graphite-type gaskets.

The event is reportable because there is a material defect in the primary coolant system that cannot be found acceptable under ASME Section XI.

(3)   <u>Degradation of Reactor Fuel Rod Cladding Identified during Fuel Sipping Operations</u>

With the plant in cold shutdown, fuel sipping operations appeared to indicate that a significant portion of cycle 2 fuel, type "LYP," had failed; i.e., 4 confirmed and 12 potential fuel leakers. The potential fuel leakers had only been sipped once before the ENS notification was made. The licensee contacted the fuel vendor for assistance onsite in evaluating this problem.

An ENS notification was made because the fuel cladding degradation was thought to be widespread. However, additional sipping operations and a subsequent evaluation by the licensee's reactor engineering department with vendor assistance concluded that no additional fuel failures had occurred; i.e., the abnormal readings associated with the potential fuel leakers was attributed to fission products trapped in the crud layer. Based on the results of the evaluation, the licensee concluded that the fuel cladding was not seriously degraded and that the event was not reportable. Consequently, after discussion with the regional office, the licensee appropriately retracted this event.

3.2.5   External Threat or Hampering

§ 50.72	§ 50.73(a)(2)(iii)
The corresponding requirement in 10 CFR 50.72 has been deleted. Refer to the plant's emergency plan regarding declaration of an emergency class.	"Any natural phenomenon or other external condition that posed an actual threat to the safety of the nuclear power plant or significantly hampered site personnel in the performance of duties necessary for the safe operation of the nuclear power plant."

An LER is required for any natural phenomenon or other external condition that poses an actual threat to the safety of the nuclear power plant or significantly hampers site personnel in the performance of duties necessary for the safe operation of the plant.

Discussion

This criterion applies only to acts of nature (e.g., tornadoes, earthquakes, fires, lightning, hurricanes, floods) and external hazards (e.g., industrial or transportation accidents). References to acts of sabotage are covered by 10 CFR 73.71, "Reporting of Safeguards

Events." Actual threats or significant hampering from internal hazards are covered by a separate criterion in 10 CFR 50.73(a)(2)(ix), as discussed in Section 3.2.10 of this report.

The phrase "actual threat to safety of the nuclear power plant" is one reporting trigger. This covers those events involving an actual threat to the plant from an external condition or natural phenomenon in which the threat or damage challenges the ability of the plant to continue to operate in a safe manner (including the orderly shutdown and maintenance of shutdown conditions).

The licensee should decide if a phenomenon or condition actually threatens the plant. For example, a minor brush fire in a remote area of the site that is quickly controlled by firefighting personnel and, as a result, did not present a threat to the plant should not be reported. However, a major forest fire, large-scale flood, or major earthquake that presents a clear threat to the plant should be reported. As another example, an industrial or transportation accident that occurs near the site, creating a plant safety concern, should be reported.

The licensee must use engineering judgment to determine if there was an actual threat. For example, with regard to tornadoes, the decision would be based on such factors as the size of the tornado and its location and path. There are no prescribed limits. In general, situations involving only monitoring by the plant's staff are not reportable, but if preventive actions are taken or if there are serious concerns, then the situation should be carefully reviewed for reportability.

Responsive actions, by themselves, do not necessarily indicate actual threats. Those that are purely precautionary, such as placement of sandbags, even though flood levels are not expected to be high enough to require sandbags, do not trigger reporting.

Section 3.2.10 of this report discusses the meaning of the phrase "significantly hampers site personnel in the performance of duties necessary for the safe operation of the plant," in the context of internal threats. A natural phenomenon or external condition may also significantly hamper personnel. If so, it is reportable under this criterion.

If a snowstorm, hurricane, or similar event significantly hampers personnel in the conduct of activities necessary for the safe operation of the plant, the event is reportable. In the case of snow, the licensee must use judgment based on the amount of snow, the extent to which personnel were hampered, the extent to which additional assistance could have been available in an emergency, the length of time the condition existed, and so forth. For example, if snow prevented shift relief for several hours, the situation would be reportable if the delay were such that site personnel were significantly hampered in the performance of duties necessary for safe operation. For example, shift personnel might exceed normal shift overtime limits, become excessively fatigued, or find it necessary to operate with fewer than the required number of watchstanders in order to allow some to rest.

Examples

(1)   Earthquake

Seismic alarms were received in the Unit 1 control room of a Southern California plant. Seismic monitors were not tripped in Units 2 or 3. The earthquake was readily felt on site. Seismic instrumentation measured less than 0.02 g lateral acceleration.

The licensee classified this as an unusual event in accordance with the emergency plan and notified the NRC via ENS per 10 CFR 50.72(a)(1)(i) within 30 minutes of the earthquake. The licensee terminated the event after walkdowns of the plant were satisfactorily completed and made an ENS update call. No LER was submitted because the event was not considered to be an actual threat.

(2)   Hurricane

A licensee in southern Florida declared an unusual event after a hurricane warning was issued by the National Hurricane Center. The hurricane was predicted to reach the site in approximately 24 hours. As part of the licensee's severe weather preparations, both operating units were taken to hot shutdown before the hurricane's predicted arrival. Offsite power to both units was lost. As the hurricane approached, wind velocity on site was measured in excess of 140 miles per hour. All personnel were withdrawn to protected safety-related structures. Extensive damage occurred on site. The unusual event was upgraded to an alert when the pressurized fire header was lost because of storm-related damage to the fire protection system water supply piping and electric pump. All safety-related equipment functioned as designed before, during, and after the storm, with the exception of two minor EDG anomalies. The licensee downgraded the alert to an unusual event once offsite power was restored and a damage assessment completed.

An ENS notification was required because the licensee declared an emergency class. An LER was required, based on the occurrence of a natural phenomenon that posed an actual threat and several other reporting criteria as well.

(3)   Fire

With the unit at 100-percent power, the control room was notified that a forest fire was burning west of the plant close to the 230-kilovolt (kV) distribution lines. Approximately 15 minutes later, voltage fluctuations were observed and then a full reactor scram occurred. The licensee determined that the offsite distribution breakers had tripped on fault, apparently from heavy smoke and heat in the vicinity of the offsite 230-kV line insulators. The other source of offsite power (i.e., the 34.5-kV lines supplying the startup transformers) was also lost. Both station EDGs received a fast start signal and load sequenced as designed. Five minutes later, offsite power was available through the startup transformer to the nonsafety-related 4,160-volt buses, but the licensee decided to maintain the vital buses on their emergency power source until the reliability of offsite power could be assured. The fire continued to burn and, although no plant structures or equipment were directly affected, the fire did approach within 70 feet of the fire pump house.

An ENS notification was required because the licensee entered the emergency plan, declaring an unusual event based on high drywell temperature and an alert based on the potential of the forest fire to further affect the plant. An LER was required, based on the occurrence of a natural phenomenon that posed an actual threat and several other reporting criteria as well.

3.2.6 System Actuation

§ 50.72(b)(2)(iv)	
"(A) Any event that results or should have resulted in emergency core cooling system (ECCS) discharge into the reactor coolant system as a result of a valid signal except when the actuation results from and is part of a pre-planned sequence during testing or reactor operation. (B) Any event or condition that results in actuation of the reactor protection system (RPS) when the reactor is critical except when the actuation results from and is part of a pre-planned sequence during testing or reactor operation."	
§ 50.72(b)(3)(iv)	**§ 50.73(a)(2)(iv)**
"(A) Any event or condition that results in valid actuation of any of the systems listed in paragraph (b)(3)(iv)(B) of this section except when the actuation results from and is part of a pre-planned sequence during testing or reactor operation.	"(A) Any event or condition that resulted in manual or automatic actuation of any of the systems listed in paragraph (a)(2)(iv)(B) of this section, except when: (1) The actuation resulted from and was part of a pre-planned sequence during testing or reactor operation; or (2) The actuation was invalid and; (i) Occurred while the system was properly removed from service; or (ii) Occurred after the safety function had been already completed.

(B) The systems to which the requirements of paragraph (b)(3)(iv)(A) of this section apply are: (1) Reactor protection system (RPS) including: reactor scram and reactor trip.[5] (2) General containment isolation signals affecting containment isolation valves in more than one system or multiple main steam isolation valves (MSIVs). (3) Emergency core cooling systems (ECCS) for pressurized water reactors (PWRs) including: high-head, intermediate-head, and low-head injection systems and the low pressure injection function of residual (decay) heat removal systems. (4) ECCS for boiling water reactors (BWRs) including: high-pressure and low-pressure core spray systems; high-pressure coolant injection system; low pressure injection unction of the residual heat removal system. (5) BWR reactor core isolation cooling system; isolation condenser system; and feedwater coolant injection system. (6) PWR auxiliary or emergency feedwater system. (7) Containment heat removal and depressurization systems, including containment spray and fan cooler systems. (8) Emergency ac electrical power systems, including: emergency diesel generators (EDGs); hydroelectric facilities used in lieu of EDGs at the Oconee Station; and BWR dedicated Division 3 EDGs.	(B) The systems to which the requirements of paragraph (a)(2)(iv)(A) of this section apply are: (1) Reactor protection system (RPS) including: reactor scram or reactor trip. (2) General containment isolation signals affecting containment isolation valves in more than one system or multiple main steam isolation valves (MSIVs). (3) Emergency core cooling systems (ECCS) for pressurized water reactors (PWRs) including: high-head, intermediate-head, and low-head injection systems and the low pressure injection function of residual (decay) heat removal systems. (4) ECCS for boiling water reactors (BWRs) including: high-pressure and low-pressure core spray systems; high pressure coolant injection system; low pressure injection function of the residual heat removal system. (5) BWR reactor core isolation cooling system; isolation condenser system; and feedwater coolant injection system. (6) PWR auxiliary or emergency feedwater system. (7) Containment heat removal and depressurization systems, including containment spray and fan cooler systems. (8) Emergency ac electrical power systems, including: emergency diesel generators (EDGs); hydroelectric facilities used in lieu of EDGs at the Oconee Station; and BWR dedicated Division 3 EDGs. (9) Emergency service water systems that do not normally run and that serve as ultimate heat sinks.
_____ [5] Actuation of the RPS when the reactor is critical is reportable under paragraph (b)(2)(iv) of this section.	

Discussion

An event that results or should have resulted in a discharge of the ECCS into the RCS as a result of a valid signal, or an event involving a critical scram, is reportable under 10 CFR 50.72(b)(2)(iv) (a 4-hour report) unless the actuation resulted from and was part of a preplanned sequence.

A valid actuation of any of the systems named in 10 CFR 50.72(b)(3)(iv)(B) is reportable under 10 CFR 50.72(b)(3)(iv)(A) (an 8-hour report) unless the actuation resulted from and was part of a preplanned sequence during testing or reactor operation.

A system actuation should be apparent at the time of occurrence. Therefore, if all events are reported properly, it is expected that all reports under 10 CFR 50.72 are as a result of an on-going condition.

An actuation of any of the systems named in 10 CFR 50.73(a)(2)(iv)(B) is reportable under 10 CFR 50.73(a)(2)(iv)(A) (a 60-day report) unless the actuation resulted from and was part of a preplanned sequence during testing or reactor operation or the actuation was invalid and occurred while the system was properly removed from service or occurred after the safety function had been already completed. As indicated in 10 CFR 50.73(a)(1), in the case of an invalid actuation reported under 10 CFR 50.73(a)(2)(iv)(A) other than actuation of the RPS when the reactor is critical, the licensee may, at its option, provide a telephone notification to the NRC Operations Center within 60 days after discovery of the event instead of submitting a written LER. In these cases, the telephone report—

(1)   Is not considered an LER.
(2)   Should identify that the report is being made under 10 CFR 50.73(a)(2)(iv)(A).
(3)   Should provide the following information:

    (a)   the specific train(s) and system(s) that were actuated
    (b)   whether each train actuation was complete or partial
    (c)   whether or not the system started and functioned successfully

These paragraphs require events to be reported whenever one of the specified systems actuates either manually or automatically. They are based on the premise that these systems are provided to mitigate the consequences of a significant event and, therefore, (1) they should work properly when called upon, and (2) they should not be challenged frequently or unnecessarily. The Commission is interested in both events in which a system was needed to mitigate the consequences of an event (whether or not the equipment performed properly) and events in which a system actuated unnecessarily.

Events involving ECCS discharge to the vessel are generally more serious than actuations without discharge to the vessel. Therefore, this reporting criterion is a 4-hour report. Valid signals that should have resulted in a discharge of the ECCS into the RCS but did not due to some component that had failed or an operator action that was taken are reportable under 10 CFR 50.72(b)(2)(iv). For example, if a valid ECCS signal was generated by plant conditions and the operator put all ECCS pumps in pull-to-lock position, although no ECCS discharge occurred, the event is reportable under 10 CFR 50.72(b)(2)(iv).

Actuations that need not be reported are those initiated for reasons other than to mitigate the consequences of an event (e.g., at the discretion of the licensee as part of a preplanned procedure).

The intent is to require reporting of the actuation of systems that mitigate the consequences of significant events. Usually, the staff would not consider this to include single-component actuations because single components of complex systems, by themselves, usually do not mitigate the consequences of significant events. However, in some cases a component would be sufficient to mitigate the event (i.e., perform the safety function) and its actuation would, therefore, be reportable. This position is consistent with the statement that the reporting requirement is based on the premise that these systems are provided to mitigate the consequences of a significant event.

Single trains do mitigate the consequences of events, and, thus, train level actuations are reportable.

In this regard, the staff considers actuation of an EDG to be actuation of a train—not actuation of a single component—because an EDG mitigates the event (performs the safety function). (See Example 3 below.)

The staff also considers intentional manual actions, in which one or more system components are actuated in response to actual plant conditions resulting from equipment failure or human error, to be reportable because such actions would usually mitigate the consequences of a significant event. This position is consistent with the statement that the Commission is interested in events in which a system was needed to mitigate the consequences of the event. For example, starting a safety injection (SI) pump in response to a rapidly decreasing pressurizer level or starting high-pressure coolant injection (HPCI) in response to a loss of feedwater would be reportable. However, shifting alignment of makeup pumps or closing a containment isolation valve for normal operational purposes would not be reportable.

Actuation of multichannel actuation systems is defined as actuation of enough channels to complete the minimum actuation logic. Therefore, single-channel actuations, whether caused by failures or otherwise, are not reportable if they do not complete the minimum actuation logic. Note, however, that if only a single logic channel actuates when, in fact, the system should have actuated in response to plant parameters, this would be reportable under these paragraphs as well as under 10 CFR 50.72(b)(3)(v) and 10 CFR 50.73(a)(2)(v) ("event or condition that could have prevented the fulfillment of the safety function of....").

With regard to preplanned actuations, operation of a system as part of a planned test or operational evolution need not be reported. Preplanned actuations are those that are expected to actually occur due to preplanned activities covered by procedures. Such actuations are those for which a procedural step or other appropriate documentation indicates that the specific actuation is actually expected to occur. Control room personnel are aware of the specific signal generation before its occurrence or indication in the control room. However, if, during the test or evolution, the system actuates in a way that is not part of the planned evolution, that actuation should be reported. For example, if the normal reactor shutdown procedure requires that the control rods be inserted by a manual reactor scram, the reactor scram need not be reported. However, if unanticipated conditions develop during the shutdown that cause an automatic reactor scram, such a reactor scram should be reported. The fact that the safety analysis assumes that a system will actuate automatically during an event does not eliminate the need to report that actuation. Actuations that need not be reported are those initiated for reasons other than to mitigate the consequences of an event (e.g., at the discretion of the licensee as part of a planned evolution).

Note that, if an operator were to manually scram the reactor in anticipation of receiving an automatic reactor scram, this would be reportable just as the automatic scram would be reportable.

Valid actuations are those actuations that result from valid signals or from intentional manual initiation, unless it is part of a preplanned test. Valid signals are those signals that are initiated in response to actual plant conditions or parameters satisfying the requirements for initiation of the system. They do not include those that are the result of other signals. Invalid actuations are, by definition, those that do not meet the criteria for being valid. Thus, invalid actuations include actuations that are not the result of valid signals and are not intentional manual actuations.

Except for critical scrams, invalid actuations are not reportable by telephone under 10 CFR 50.72. In addition, invalid actuations are not reportable under 10 CFR 50.73 in any of the following circumstances:

(1) The invalid actuation occurred when the system was already properly removed from service. This means that all requirements of plant procedures for removing equipment from service have been met. It includes required clearance documentation, equipment and control board tagging, and properly positioned valves and power supply breakers.

(2) The invalid actuation occurred after the safety function had already been completed. An example would be RPS actuation after the control rods have already been inserted into the core.

If an invalid actuation reveals a defect in the system so that the system failed or would fail to perform its intended function, the event continues to be reportable under other requirements of 10 CFR 50.72 and 50.73. When invalid actuations excluded by the conditions described above occur as part of a reportable event, they should be described as part of the reportable event in order to provide a complete, accurate, and thorough description of the event.

Examples

(1)   Reactor Protection System Actuation

•   The licensee was placing the RHR system in its shutdown cooling mode while the plant was in hot shutdown. The BWR vessel level decreased for unknown reasons, causing RPS scram and Group III primary containment isolation signals, as designed. All control rods had been previously inserted and all Group III isolation valves had been manually isolated. The licensee isolated RHR to stop the decrease in reactor vessel level.

An ENS notification and an LER are both required because, although the systems' safety functions had already been completed, the RPS scram and primary containment isolation signals were valid and the actuations were not part of the planned procedure. The automatic signals were valid because they were generated from the sensor by measurement of an actual physical system parameter that was at its setpoint.

- With the BWR defueled, an invalid signal actuated the RPS. There was no component operation because the control rod drive system had been properly removed from service. This event is not reportable because (1) the RPS signal was invalid, and (2) the system had been properly removed from service.

- At a BWR, both recirculation pumps tripped as a result of a breaker problem. This placed the plant in a condition in which BWRs are typically scrammed to avoid potential power/flow oscillations. At this plant, for this condition, a written off-normal procedure required the plant operations staff to scram the reactor. The plant staff performed a reactor scram, which was uncomplicated.

  This event is reportable as a manual RPS actuation. Even though the reactor scram was in response to an existing written procedure, this event does not involve a preplanned sequence because the loss of recirculation pumps and the resultant off-normal procedure entry were event driven, not preplanned. Both an ENS notification and an LER are required. In this case, the licensee initially retracted the ENS notification, believing that the event was not reportable. After staff review and further discussion, it was agreed that the event is reportable for the reasons discussed above.

(2)   Boiling Water Reactor Control Rod Block Monitor Actuation

A rod block that was part of the planned startup procedure occurred from the rod block monitor, which, at this plant, is classified as a portion of the RPS.

This event is not reportable because it occurred as a part of a preplanned startup procedure that specified that certain rod blocks were expected to occur.

(3)   Emergency Diesel Generator Starts

- An EDG automatically started when a technician inadvertently caused a short circuit that de-energized an essential bus during a calibration. The actuation was valid because an essential bus was de-energized. The event is reportable because the EDG autostart was not identified at the step in the calibration procedure being used.

- After an automatic EDG start, and for unknown reasons, the emergency bus feeder breaker from the EDG did not close when power was lost on the bus. The event is reportable because the actuation logic for the EDG start was completed, even though the EDG did not power the safety buses.

(4)   Preplanned Manual Scram

During a normal reactor shutdown, the reactor shutdown procedure required that reactor power be reduced to a low power, at which point the control rods were to be inserted by a manual reactor scram. The rods were manually scrammed.

This event is not reportable because the manual scram resulted from and was, by procedure, part of a preplanned sequence of reactor operation. However, if conditions develop during the process of shutting down that require an unplanned reactor scram, the RPS actuation (whether manually or automatically produced) is reportable.

(5)   Actuation of Wrong Component during Testing

During surveillance testing of the MSIVs, an operator incorrectly closed MSIV D when the procedure specified closing MSIV C.

This event is not reportable because the event is an inadvertent actuation of a single component rather than a train-level actuation (and the purpose of the actuation was not to mitigate the consequences of an event).

(6)   Reactor Water Cleanup Isolation

A Reactor Water Cleanup (RWCU) primary containment isolation occurred on pressurization between the RWCU suction containment isolation valves, as designed to isolate a pipe break. It is a valid signal because this is the safety function of the containment isolation system. Regardless, the event is not reportable because the signal did not affect containment isolation valves in multiple systems.

(7)   Manual Actuation of Component in Response to Actual Plant Condition

At a PWR, maintenance personnel inadvertently pulled an instrument line out of a compression fitting connection at a pressure transmitter. The resultant RCS leak was estimated at between 70 and 80 gallons per minute. Charging flow increased due to automatic control system action. The operations staff recognized the symptoms of an RCS leak and entered the appropriate off-normal procedure. The procedure directed the operations staff to start a second charging pump, and flow was manually increased to raise pressurizer level. Based on the response of the pressurizer level, the operations staff determined that a reactor scram and SI were not necessary. Maintenance personnel still at the transmitter closed the instrument block and root valves, terminating the event.

The staff considers the manual start of the charging pump (which also serves as an ECCS pump, but with a different valve lineup) in response to dropping pressurizer level to be an intentional manual actuation in response to equipment failure or human error and reportable because it constitutes deliberate manual actuation of a single component, in response to plant conditions, to mitigate the consequences of an event. As discussed previously in this section, actuations that need not be reported are those that are initiated for reasons other than to mitigate the consequences of an event (e.g., at the discretion of the licensee as part of a planned procedure or evolution).

(8) Actuation during Maintenance Activity

At a BWR, a maintenance activity was underway involving placement of a jumper to avoid unintended actuations. The maintenance staff recognized that there was a high potential for a loss of contact with the jumper and consequent actuation. This potential was explicitly stated in the maintenance work request and on a risk evaluation sheet. The operating staff was briefed on the potential actuations prior to start of work. During the event, a loss of continuity did occur and the actuations occurred, involving isolation, standby gas treatment start, closing of some valves in the primary containment isolation system (recirculation pump seal mini purge valve, nitrogen supply to drywell valve, and containment atmospheric monitoring valve).

The event is not reportable under 10 CFR 50.72(b)(2)(iv) or (b)(3)(iv) because the actuations were not valid. It is reportable under 10 CFR 50.73(a)(2)(iv) because the actuations were not listed as (and were not) definitely expected to occur.

3.2.7 Event or Condition that Could Have Prevented Fulfillment of a Safety Function

§ 50.72(b)(3)(v)	§ 50.73(a)(2)(v)
"Any event or condition that at the time of discovery could have prevented the fulfillment of the safety function of structures or systems that are needed to:	"Any event or condition that could have prevented the fulfillment of the safety function of structures or systems that are needed to:
(A) Shut down the reactor and maintain it in a safe shutdown condition;	(A) Shut down the reactor and maintain it in a safe shutdown condition;
(B) Remove residual heat;	(B) Remove residual heat;
(C) Control the release of radioactive material; or	(C) Control the release of radioactive material; or
(D) Mitigate the consequences of an accident."	(D) Mitigate the consequences of an accident."
§ 50.72(b)(3)(vi)	**§ 50.73(a)(2)(vi)**
"Events covered in paragraph (b)(3)(v) of this section may include one or more procedural errors, equipment failures, and/or discovery of design, analysis, fabrication, construction, and/or procedural inadequacies. However, individual component failures need not be reported pursuant to paragraph (b)(3)(v) of this section if redundant equipment in the same system was operable and available to perform the required safety function."	"Events covered in paragraph (a)(2)(v) of this section may include one or more procedural personnel errors, equipment failures, and/or discovery of design, analysis, fabrication, construction, and/or procedural inadequacies. However, individual component failures need not be reported pursuant to paragraph (a)(2)(v) of this section if redundant equipment in the same system was operable and available to perform the required safety function."

If the event or condition could have prevented fulfillment of the safety function at the time of discovery, an ENS notification and an LER are required. If it could have prevented fulfillment of the safety function at any time within 3 years of the date of discovery, but not at the time of discovery, only an LER is required. If the event or condition could have prevented fulfillment of the safety function at the time of discovery, and if it is not reported under 10 CFR 50.72(a), (b)(1), or (b)(2), an ENS notification is required under 10 CFR 50.72(b)(3).

Discussion

This criterion is based on the assumption that safety-related SSCs are intended to mitigate the consequences of an accident. SSCs within scope include only safety-related SSCs required by the TS to be operable that are intended to mitigate the consequences of an accident as discussed in Chapters 6 and 15 of the Final Safety Analysis Report (or equivalent chapters). Accidents are identified as events of moderate frequency, infrequent incidents, or limiting faults as discussed in Regulatory Guide 1.70, "Standard Format and Content of Safety Analysis Reports for Nuclear Power Plants (LWR Edition)" (or equivalent classifications of the three types of events). The American Nuclear Society (ANS) categorizes these events as Condition II, III, and IV type events.

The level of judgment for reporting an event or condition under this criterion is a reasonable expectation of preventing fulfillment of a safety function. In the discussions that follow, many of which are taken from previous NUREG guidance, several different expressions, such as "would have," "could have," "alone could have," and "reasonable doubt," are used to characterize this standard. In the staff's view, all of these should be judged on the basis of a reasonable expectation of preventing fulfillment of the safety function. A SSC that has been declared inoperable is one in which the SSC capability is degraded to a point where it cannot perform with reasonable expectation or reliability. These criteria cover an event or condition in which scoped in SSCs could have failed to perform their intended function because of one or more personnel errors, including procedure violations; equipment failures; inadequate maintenance; or design, analysis, fabrication, equipment qualification, construction, or procedural deficiencies and no redundant equipment in the same system was operable.

As a result, for SSCs within the scope of this criterion, a report is required when 1) there is a determination that the SSC is inoperable in a required mode or other specified condition in the TS Applicability, 2) the inoperability is due to one or more personnel errors, including procedure violations; equipment failures; inadequate maintenance; or design, analysis, fabrication, equipment qualification, construction, or procedural deficiencies, and 3) no redundant equipment in the same system was operable. For guidance on determining whether a SSC is operable, see RIS 2005-20, Revision 1. Operable but nonconforming or degraded conditions are not considered reportable under this criterion.

As a result, reports are not required when systems are declared inoperable as part of a planned evolution for maintenance or surveillance testing when done in accordance with an approved procedure and the plant's TS (unless a condition is discovered that would have resulted in the system being declared inoperable). In addition, unless a condition is discovered that would have resulted in the system being declared inoperable, reports are not required when systems are declared inoperable solely as a result of Required Actions for which the bases is the assumption of an additional random single failure (i.e. Westinghouse STS, Revision 4, LCO 3.8.1, "AC Sources – Operating," Required Actions A.2, B.2, or C.1 (ADAMS Accession No.ML12100A222)).

The event must be reported regardless of whether or not an alternate safety system could have been used to perform the safety function. For example, if the onsite power system was declared inoperable due to equipment failures, the event would be reportable, even if the offsite power system remained operable.

For systems that include three or more trains, the inoperability of two or more trains should be reported if, in the judgment of the licensee, the remaining operable trains could not mitigate the consequences of an accident.

There are a limited number of single-train systems that perform safety functions (e.g., the HPCI system in BWRs). For such systems, inoperability of the single train is reportable even though the plant TS may allow such a condition to exist for a limited time.

If the retraction or cancellation of a report under this criterion is due to a revised operability determination, the retraction or cancellation should discuss the basis for why the operability determination was revised, and why it is believed that system operability was never lost (i.e., in lieu of the initial determination).

Examples

SINGLE-TRAIN SYSTEMS

(1)   Failure of a Single-Train System Preventing Accident Mitigation and Residual Heat Removal

When the licensee was preparing to run a surveillance test, an HPCI flow controller was found to be inoperable; therefore, the licensee declared the HPCI system inoperable. The plant entered a TS requiring that the automatic depressurization, low-pressure coolant injection, core spray, and isolation condenser systems remain operable during the 7-day LCO or the plant would have to be shut down.

The licensee made an ENS notification within 28 minutes and a followup call after the amplifier on the HPCI flow transmitter was fixed and the HPCI returned to operability. As discussed above, the loss of a single-train safety system such as BWR HPCI is reportable.

(2)   Failure of a Single-Train Nonsafety System

Question: If reactor core isolation cooling (RCIC) is not a "safety system" in that no credit for its operation is taken in the safety analysis, are failures and unavailability of this system reportable?

Answer: If the plant's safety analysis considered RCIC as a system needed to mitigate a rod ejection accident and it is included in the TS, then its failure is reportable under this criterion; otherwise, it is not reportable under this section of the rule.

(3)   Failure of a Single-Train Environmental System

Question: There are a number of environmental systems in a plant dealing with such things as low-level waste (e.g., gaseous radwaste tanks). Many of these systems are not required to meet the single failure criterion, so a single failure results in the loss of function of the system. Are all of these systems covered within the scope of the LER rule?

Answer: Such systems would be within scope if they are safety-related systems retained in the TS that are intended to mitigate the consequences of an accident.

LOSS OF TWO TRAINS

(4)   Loss of Onsite Emergency Power by Multiple Equipment Inoperability and Unavailability

During refueling, one EDG in a two-train system was out of service for maintenance. The second EDG was declared inoperable when it failed its surveillance test.

An ENS notification is required and an LER is required. As addressed in the discussion section above, loss of the onsite power system is reportable under this criterion.

(5)   Procedure Error Prevents Reactor Shutdown Function

The unit was in Mode 5 (cold and depressurized, before initial criticality) and a postmodification test was in progress on the train A RPS, when the operator observed that both train A and B source range detectors were disabled. During postmodification testing on the train A RPS, instrumentation personnel placed the train B input error inhibit switch in the inhibit position. With both trains' input error inhibit switches in the inhibit position, source range detector voltage was disabled. The input error inhibit switch was immediately returned to the normal position, and a caution was added to appropriate plant instructions.

This event is reportable because disabling the source range detectors could have prevented fulfillment of the safety function to shut down the reactor.

(6)   Failure of the Overpressurization Mitigation System

The RCS was overpressurized on two occasions during startup following a refueling outage because the overpressure mitigation system (OMS) failed to operate. The reason that the OMS failed to operate was that one train was out of service for maintenance, a pressure transmitter was isolated, and a summator failed in the actuation circuit on the other train.

The event is reportable because the OMS failed to perform its safety function.

(7)   Loss of Saltwater Cooling System and Flooding in Saltwater Pump Bay

During maintenance activities on the south saltwater pump, the licensee was removing the pump internals from the casing when flooding of the pump area occurred. The north saltwater pump was secured to prevent pump damage.

The event is reportable because of the failure of the saltwater cooling system, which is the ultimate heat sink for the facility, to perform its safety function.

(8)   Maintenance Affecting Two Trains

Question: Some clarification is needed for events or conditions that "could have" prevented the fulfillment of a system safety function.

Answer: With regard to maintenance problems, events or conditions generally involve operator actions and/or component failures that could have prevented the functioning of a safety system. For example, assume that a surveillance test is run on a standby pump and it seizes. The pump is disassembled and found to contain the wrong lubricant. The redundant pump is disassembled and it also has the same wrong lubricant. Thus, it is reasonable to assume that the second pump would have failed if it had been challenged. However, the second pump and, therefore, the system did not actually fail because the second pump was never challenged. Thus, in this case, because of the use of the wrong lubricant, the system "could have" or "would have" failed.

LOSS OF ONE TRAIN

(9)   Contaminated Hydraulic Fluid Degrades Main Steam Isolation Valve Operation

Situation: During a routine shutdown, the operator noted that the #11 MSIV closing time appeared to be excessive. A subsequent test revealed the #11 MSIV to shut within the required time; however, the #12 MSIV closing time exceeded the maximum at 7.4 seconds. Contamination of the hydraulic fluid in the valve actuation system had caused the system's check valves to stick and delay the transmission of hydraulic pressure to the actuator. The licensee will purchase three more filters, providing supplemental filtering for each MSIV. Finer filters will be used in pump suction filters to remove the fine contaminants. The #12 MSIV was repaired and returned to service. Because the valves were not required for operation at the time of discovery, the safety of the public was not affected.

Comments: The event is reportable under 10 CFR 50.73(a)(2)(v) because the condition could have prevented fulfillment of a safety function. The event is not reportable under 10 CFR 50.72(b)(3)(v) because, at the time of discovery, the plant was shut down and the MSIVs were not required to be operable.

(10)   Emergency Diesel Generator Lube Oil Fire Hazard

Situation: While the licensee was performing a routine surveillance test of the EDG, a small fire started due to lubricating oil leakage from the exhaust manifold. The manufacturer reviewed the incident and determined that the oil was accumulating in the exhaust manifold due to leakage originating from above the upper pistons of this vertically opposed piston engine. The oil remaining above the upper pistons after shutdown leaked slowly down past the piston rings, into the combustion space, past the lower piston rings, through the exhaust ports, and into the exhaust manifolds. The exhaust manifolds became pressurized during the subsequent startup, which forced the oil out through leaks in the exhaust manifold gaskets where it was ignited. Similar events occurred previously at this plant. In these previous cases, fuel oil accumulated in the exhaust manifold due to extended operation under "no load" conditions. Operation under loaded conditions was therefore required before shutdown in order to burn off any accumulated oil.

Comments: The event is not reportable if the fire did not pose a threat to the plant (e.g., it did not significantly hamper site personnel per 10 CFR 50.73(a)(2)(x)). The event would be reportable if it demonstrated a design, procedural, or equipment deficiency that could have prevented the fulfillment of a safety function (i.e., if the redundant diesels are of similar design and, therefore, susceptible to the same problem) (10 CFR 50.73(a)(2)(vi)).

(11)   Single Failures

Question: Suppose you have one pump in a cooling water system (e.g., chilled water) supplying water to both trains of a safety system, but there is another pump in standby; is the loss of the one operating pump reportable?

Answer: No. Single, independent (i.e., random) component failures are not reportable if the redundant component in the same system did or would have fulfilled the safety function. However, if such failures have generic implications, then an LER is to be submitted.

(12)   Generic Setpoint Drift

·   Situation: With the plant in steady-state power operation and while performing a main steamline pressure instrument functional test and calibration, the licensee found a switch to actuate at 853 pounds per square inch, gauge. The TS limit is 825+15. The redundant switches were operable. The cause of the occurrence was setpoint drift. The switch was recalibrated and tested successfully per HNP-2-5279, "Barksdale Pressure Switch Calibration," and returned to service. This is a repetitive event as reported in one previous LER. A generic review revealed that this type of switch is used on other safety systems and that this type of switch is subject to drift. An investigation will continue as to why these switches drift, and, if necessary, they will be replaced.

Comments: The event is not reportable due to the drift of a single pressure switch. The event is reportable if it is indicative of a generic and/or repetitive problem with this type of switch, which is used in several safety systems (10 CFR 50.73(a)(2)(vi) or (vii)).

·   Question: Are setpoint drift problems with a particular switch to be reported if they are experienced more than once?

Answer: The independent failure (e.g., excessive setpoint drift) of a single pressure switch is not reportable unless it could have caused a system to fail to fulfill its safety function or is indicative of a generic problem that could have resulted in the failure of more than one switch and thereby cause one or more systems to fail to fulfill their safety functions.

(13)   Maintenance Affecting Only One Train

Question: Suppose the wrong lubricant was installed in one pump, but the pump in the other train was correctly lubricated. Is this reportable?

Answer: Engineering judgment is required to decide if the lubricant could have been used on the other pump, and, therefore, the system function would have been lost. If the procedure called for testing of the first pump before maintenance was performed on the second pump, and testing clearly identified the error, then the error would not be reportable. However, if the procedure called for the wrong lubricant and eventually both pumps would have been improperly lubricated, and the problem was only discovered when the first pump was actually challenged and failed, then the error would be reportable.

OTHER CONDITIONS

(14)   Conditions Observed While System Out of Service

Question: Suppose that, during shutdown, we are doing maintenance on both SI pumps, which are not required to be operational. Is this reportable? While shut down, suppose I identify or observe something that would cause the SI pumps not to be operational at power. Is this reportable?

Answer: Removing both SI pumps from service to do maintenance is not reportable if the resulting system configuration is not prohibited by the plant's TS. However, if a situation is discovered during maintenance that could have caused both pumps to fail (e.g., they are both improperly lubricated), then that condition is reportable in an LER even though the pumps were not required to be operational at the time that the condition was discovered.

As another example, suppose the scram breakers were tested during shutdown conditions, and it was found that opening times for more than one breaker were in excess of those specified, or that undervoltage trip attachments were inoperative. Such potential generic problems are reportable in an LER.

(15)   Emergency Diesel Generator Bearing Problems

During the annual inspection of one standby EDG, the lower crankshaft thrust bearing and adjacent main bearing were found to be wiped on the journal surface. The thrust bearing was also found to have a small crack from the main oil supply line across the journal surface to the thrust surface. Inspection of the second, redundant standby EDG revealed similar problems. It was judged that extended operation without corrective action could have resulted in bearing failure.

The event is reportable because there was reasonable doubt that the diesels would have completed an extended run under load, as required, if called upon.

(16)   Multiple Control Rod Failures

There have been cases in which licensees have erroneously concluded that sequentially discovered failures of systems or components occurring during planned testing are not reportable. The NRC identified this situation as a generic concern on April 3, 1985, in Information Notice (IN) 85-27, "Notifications to the NRC Operations Center and Reporting Events in Licensee Event Reports," regarding the reportability of multiple events in accordance with 10 CFR 50.72(b)(3)(v) and 10 CFR 50.73(a)(2)(v) (event or condition that could have prevented fulfillment of a safety function).

IN 85-27 described multiple failures of an RPS during control rod insertion testing of a reactor at power. One of the control rods stuck. Subsequent testing identified 3 additional rods that would not insert (scram) into the core and 11 control rods that had an initial hesitation before insertion. The licensee considered each failure as a single random failure; thus, each was determined not to be reportable. Subsequent assessments indicated that the instrument air system, which was to be oil free, was contaminated with oil that was causing the scram solenoid valves to fail. Although the failure of a single rod to insert may not cause a reasonable doubt about the ability of other rods to insert, the failure of more than one rod does cause a reasonable doubt.

As indicated in IN 85-27, multiple failures of redundant components of a safety system are sufficient reason to expect that the failure mechanism, even though not known, could have prevented the fulfillment of the safety function.

(17)   Potential Loss of High-Pressure Coolant Injection

During normal refueling leak testing of the upstream containment isolation check valve on the HPCI steam exhaust, the disc of the noncontainment isolation check valve was found to be lodged in downstream piping. This might have prevented HPCI from functioning if the disc had blocked the line. The event was caused by fatigue failure of a disc pin.

Following evaluation of the condition, the event was determined to be reportable because the HPCI could have been prevented from performing its safety function if the disc had blocked the line. In addition, the event is reportable if the fatigue failure is indicative of a common mode failure.

(18)   Operator Inaction or Wrong Action

Question: In some systems used to control the release of radioactivity, a detector controls certain equipment. In other systems, a monitor is present and the operator is required to initiate action under certain conditions. The operator is not "wired" in. Are failures of the operator to act reportable?

Answer: Yes. The operator may be viewed as a "component" that is an integral, and frequently essential, part of a "system." Thus, if an event or condition meets the reporting criterion, it is to be reported regardless of the initiating cause.

(19)   Results of Analysis

Question: A number of criteria indicate that they apply to actual situations only and not to potential situations identified as a result of analysis; yet, other criteria address "could have." When do the results of analysis have to be reported?

Answer: The results need only be reported if the applicable criterion requires the reporting of conditions that "could have" caused a problem. However, others have a need to know about potential problems that are not reportable; thus, such items may be reported as a voluntary LER.

(20)   System Interactions

Question: Utilities are not required to analyze for system interactions, yet the rule requires the reporting of events that "could have" happened but did not. Are we to initiate a design activity to determine "could have" system interactions?

Answer: No. Report system interactions that you find as a result of ongoing, routine activities (e.g., the analysis of operating events).

3.2.8 Common Cause Inoperability of Independent Trains or Channels

§ 50.72	§ 50.73(a)(2)(vii)
There is no corresponding requirement in 10 CFR 50.72.	"Any event where a single cause or condition caused at least one independent train or channel to become inoperable in multiple systems or two independent trains or channels to become inoperable in a single system designed to: (A) Shut down the reactor and maintain it in a safe shutdown condition; (B) Remove residual heat; (C) Control the release of radioactive material; or (D) Mitigate the consequences of an accident."

An LER is required for a common cause inoperability of independent trains or channels.

Discussion

This criterion requires those events to be reported in which a single cause or condition caused independent trains or channels to become inoperable. Common causes may include such factors as high ambient temperatures, heatup from energization, inadequate preventive maintenance, oil contamination of air systems, incorrect lubrication, use of nonqualified components, or manufacturing or design flaws. The event is reportable if the independent trains or channels were inoperable at the same time, regardless of whether or not they were discovered at the same time. (Example 2 below illustrates a case in which the second failure was discovered 3 days later than the first.)

An event or failure that results in or involves the failure of independent portions of more than one train or channel in the same or different systems is reportable. For example, if a cause or condition caused components in train A and B of a single system to become inoperable, even if additional trains (e.g., train C) were still available, the event must be reported. In addition, if the cause or condition caused components in train A of one system and in train B of another system (i.e., the train that is assumed in the safety analysis to be independent) to become inoperable, the event must be reported. However, if a cause or condition caused components in train A of one system and train A of another system (i.e., trains that are not assumed in the safety analysis to be independent), the event need not be reported unless it meets one or more of the other reporting criteria.

Trains or channels, for reportability purposes, are defined as those redundant, independent trains or channels designed to provide protection against single failures. Many engineered safety systems containing active components are designed with at least a two-train system. Each independent train in a two-train system can normally satisfy all of the safety system requirements to safely shut down the plant or satisfy those criteria that have to be met following an accident.

This criterion does not include those cases in which one train of a system or a component was removed from service as part of a planned evolution, in accordance with an approved procedure, and in accordance with the plant's TS. For example, if the licensee removes part of a system from service to perform maintenance, and the TS permit the resulting configuration, and the system or component is returned to service within the time limit specified in the TS, the action need not be reported under this paragraph. However, if, while the train or component is out of service, the licensee identifies a condition that could have prevented the whole system from performing its intended function (e.g., the licensee finds a set of relays that is wired incorrectly), that condition must be reported.

Analysis of events reported under this part of the rule may identify previously unrecognized common cause (or dependent) failures and system interactions. Such failures can be simultaneous failures that occur because of a single initiating cause (i.e., the single cause or mechanism serves as a common input to the failures), or the failures can be sequential (i.e., cascading failures), such as the case in which a single component failure results in the failure of one or more additional components.

Examples

(1)   Incorrect Lubrication Degrades Main Steam Isolation Valve Operation

During monthly operability tests, the licensee found that the Unit 2B inboard MSIV did not stroke properly as a result of a solenoid-operated valve failure. Both units were shut down from 100-percent power, and the solenoid-operated valves piloting all 16 MSIVs were inspected. The licensee found that the solenoid-operated valves on all 16 MSIVs were damaged. The three-way and four-way valves and solenoid pilot valves on all 16 MSIVs had a hardened, sticky substance in their ports and on their O-rings. As a result, motion of all the solenoid-operated valves was impaired, resulting in instrument air leakage and the inability to operate all of the MSIVs satisfactorily. The licensee also examined unused spares in the warehouse and found that the lubricant had dried out in those valves, leaving a residue. Several of the warehouse spares were bench tested. They were found to be degraded and also leaked. The root cause of the event was use of an incorrect lubricant.

The event is reportable (1) because a single cause or condition caused multiple independent trains of the main steam isolation system (a system designed to control the release of radioactive material and mitigate the consequences of an accident) to become inoperable (10 CFR 50.73(a)(2)(vii)(C) and (D)), and (2) because a single condition could have prevented fulfillment of a safety function (10 CFR 50.73(a)(2)(v)).

(2)   Marine Growth Causing Emergency Service Water To Become Inoperable (Common Mode Failure Mechanism)

With Unit 1 at 74-percent power and Unit 2 at 100-percent power, emergency service water (ESW) pump 1A was declared inoperable because its flow rate was too low to meet acceptance criteria. Three days later, with both units at the same conditions, ESW pump 1C was declared inoperable for the same reason. The ESW pumps are the source of water from the intake canal during a design-basis accident. In both cases, the cause was marine growth of hydroids and barnacles on the impeller and suction of the pumps. Following maintenance, both pumps passed their performance tests and were placed in service. Pump testing frequency was increased to more closely monitor pump performance.

This event is reportable because a single cause or condition caused two independent trains to become inoperable in a single system designed to mitigate the consequences of an accident (10 CFR 50.73(a)(2)(vii)(D)).

(3)   Testing Indicated Several Inoperable Snubbers

The licensee found 11 inoperable snubbers during periodic testing. All of the snubbers failed to lock up in tension and/or compression. These failures did not render their respective systems inoperable but did render trains inoperable. Improper lockup settings and/or excessive seal bypass caused these snubbers to malfunction. These snubbers were designed for low-probability seismic events. Numerous previous similar events have been reported by this licensee.

This condition is reportable because the condition indicated a generic common mode problem that caused numerous multiple independent trains in one or more safety systems to become inoperable. The potential existed for numerous snubbers in several systems to fail following a seismic event, rendering several trains inoperable (10 CFR 50.73(a)(2)(vii)).

(4)   Stuck High-Pressure Injection System Check Valves as a Result of Corroded Flappers

The licensee reported that check valves in three of four high-pressure injection (HPI) lines were stuck closed. The unit had been shut down for refueling and maintenance.

A special test of the check valves revealed that three stop check valves failed to open when a differential pressure up to the capacity of the pump was applied. Further review showed that the common cause of valve failure was the flappers corroding shut.

The event is reportable because a single cause or condition caused at least two independent trains of the HPI system to become inoperable. This system is designed to remove residual heat and mitigate the consequences of an accident. The condition is therefore reportable under 10 CFR 50.73(a)(2)(vii)(B) and (D) (common cause failure in systems designed to remove residual heat and mitigate accidents).

3.2.9 Radioactive Release

§ 50.72	§ 50.73(a)(2)(viii)
The corresponding requirement in 10 CFR 50.72 has been deleted.	"(A) Any airborne radioactivity release that, when averaged over a time period of 1 hour, resulted in airborne radionuclide concentrations in an unrestricted area that exceeded 20 times the applicable concentration limits specified in Appendix B to Part 20, Table 2, Column 1.
Refer to the plant's **emergency plan** regarding declaration of an emergency class.	
Refer to **10 CFR 50.72(b)(2)(xi)** below regarding a news release or notification of	(B) Any liquid effluent release that, when

| another agency.

Refer to **10 CFR 20.2202** regarding events reportable under that section.	averaged over a time period of 1 hour, exceeds 20 times the applicable concentrations specified in Appendix B to Part 20, Table 2, Column 2, at the point of entry into the receiving waters (i.e., unrestricted area) for all radionuclides except tritium and dissolved noble gases."

An LER is required for a release as defined in the rules.

Discussion

Although similar to 10 CFR 20.2202 and 20.2203, these criteria place a lower threshold for reporting events at commercial power reactors because the significance of the breakdown of the licensee's program that allowed such a release is the primary concern, rather than the significance of the effect of the actual release. In contrast, however, the time limit for reporting under 10 CFR 20.2202 and 20.2203 is more restrictive.

For a release that took less than 1 hour, normalize the release to 1 hour (e.g., if the release lasted 15 minutes, divide by 4). For releases that lasted more than 1 hour, use the highest release for any continuous 60-minute period (i.e., comparable to a moving average).

Annual average meteorological data should be used for determining offsite airborne concentrations of radioactivity to maintain consistency with the TS for reportability thresholds.

The location used as the point of release for calculation purposes should be determined using the expanded definition of an unrestricted area as specified in NUREG-0133, "Preparation of Radiological Effluent Technical Specifications for Nuclear Power Plants," issued October 1978, to maintain consistency with the TS.

Examples

(1) Unmonitored Release of Contaminated Steam through Auxiliary Boiler Atmospheric Vent

An unmonitored release of contaminated steam resulted from a combination of a tube leak, improper venting of an auxiliary boiler system, and inadequate procedures. This combination resulted in a release path from a liquid waste concentrator to the atmosphere via the auxiliary boiler system steam drum vent.

Because of rain at the site, the steam release to the atmosphere was condensed and deposited onto plant buildings and yard areas. This contamination was washed via a storm drain into a lake. The release was later confirmed to be 2.6×10^{-5} microcuries per milliliter (μCi/ml) of caesium-137 at the point of entry into the receiving water.

An LER for a liquid radioactive material release is required because the unmonitored release exceeded 20 times the applicable concentrations specified in Table 2, Column 2 of Appendix B, "Annual Limits on Intake (ALIs) and Derived Air Concentrations (DACs) of Radionuclides for Occupational Exposure; Effluent Concentrations; Concentrations for

Release to Sewerage," to 10 CFR Part 20, "Standards for Protection against Radiation," averaged over 1 hour at the site boundary.

(2)   Unplanned Gaseous Release

During routine scheduled maintenance on a pressure-actuated valve in the gaseous waste system, an unplanned radioactive release to the environment was detected by a main stack high radiation alarm. The release occurred when an isolation valve, required to be closed on the station tagout sheet, was inadvertently left open. This allowed radioactive gas from the waste gas decay tank to escape through a pressure gage connection that had been opened to vent the system. Operator error was the root cause of this release, with ambiguous valve tag numbers as a contributing factor. The concentration in the unrestricted area, averaged over 1 hour, was estimated by the licensee to be 1×10^{-5} µCi/ml of krypton-85 and 5×10^{-6} µCi/ml of xenon-133.

The event was reportable via an LER because the sum of the ratios of the concentration of each airborne radionuclide in the restricted area when averaged over a period of 1 hour, to its respective concentration specified in Table 2, Column 1 of Appendix B to 10 CFR Part 20, exceeds 20.

3.2.10 Internal Threat or Hampering

§ 50.72	§ 50.73(a)(2)(x)
The corresponding requirement in 10 CFR 50.72 has been deleted. Refer to the plant's **emergency plan** regarding declaration of an emergency class.	"Any event that posed an actual threat to the safety of the nuclear power plant or significantly hampered site personnel in the performance of duties necessary for the safe operation of the nuclear power plant including fires, toxic gas releases, or radioactive releases."

An LER is required for an event that poses an actual threat or causes significant hampering, as defined in the rules.

Discussion

These criteria pertain to internal threats. The criterion for external threats, 10 CFR 50.73(a)(2)(iii), is described in Section 3.2.5 of this report.

This provision requires reporting events, particularly those caused by acts of personnel, that endanger the safety of the plant or interfere with personnel in the performance of duties necessary for safe plant operations.

The licensee must exercise some judgment in reporting under this rule. For example, a small fire on site that did not endanger any plant equipment and did not and could not reasonably be expected to endanger the plant is not reportable.

The phrase "significantly hampers site personnel" applies narrowly; i.e., only to those events that significantly hamper the ability of site personnel to perform safety-related activities affecting plant safety.

In addition, the staff considers the following standards appropriate in this regard:

- The significant hampering criterion is pertinent to "the performance of duties necessary for safe operation of the nuclear power plant." One way to evaluate this is to ask if one could seal the room in question (or disable the function in question) for a substantial period of time and still operate the plant safely. For example, if a switchgear room is unavailable for a time, but it is normally not necessary to enter the room for safe operation, and no need to enter the room arises while it is unavailable, the event is not reportable under this criterion.

- Significant hampering includes hindering or interfering, provided that the interference or delay is sufficient to significantly threaten the safe operation of the plant.

- Actions such as room evacuations that are precautionary would not constitute significant hampering if the necessary actions can still be performed in a timely manner.

Plant mode may be considered in determining if there is an actual internal threat to a plant. However, licensees should not incorrectly assume that everything that happens while a plant is shut down is unimportant and not reportable.

In-plant releases must be reported if they require evacuation of rooms or buildings and, as a result, the ability of the operators to perform duties necessary for safe operation of the plant is significantly hampered.

Events such as minor spills, small gaseous waste releases, or the disturbance of contaminated particulate matter (e.g., dust) that require temporary evacuation of an individual room until the airborne concentrations decrease or until respiratory protection devices are used, are not reportable unless the ability of site personnel to perform necessary safety functions is significantly hampered.

No LER is required for precautionary evacuations of rooms and buildings that subsequent evaluation determines were not required. Even if an evacuation affects a major part of the facility, the test for reportability is whether an actual threat to plant safety occurred or whether site personnel were significantly hampered in carrying out their safety responsibilities.

In most cases, fires result in an ENS notification because there is a declaration of an emergency class, which is reportable under 10 CFR 50.72(a)(1)(i) as discussed in Section 3.1.1 of this report.[9] If there is an actual threat or significant hampering, an LER is also required. With

---

[9] As indicated in NUREG-0654, Revision 1, IN 88-64, "Reporting Fires in Nuclear Process Systems at Nuclear Power Plants," dated August 18, 1988, and Regulatory Guide 1.101, Revision 3, "Emergency Planning and Preparedness for Nuclear Power Reactors," issued August 1992 (which endorses NUMARC/NESP-007, Revision 2), a fire that lasts longer than 10 or 15 minutes or that affects plant equipment important for safe operation would result in declaration of an emergency class.

regard to control room fires, the staff generally considers a control room fire to constitute an actual threat and significant hampering.[10]

Examples

(1) <u>Fire in Refueling Bridge</u>

Question: If we have a fire in the refueling bridge and we are not moving fuel, would the fire be reportable?

Answer: No. If the plant is not moving fuel and the fire does not otherwise threaten other safety equipment and does not hamper site personnel, the fire is not reportable. If the plant is moving fuel, the fire is reportable.

(2) <u>Fire in Reactor Building</u>

Question: If we have a fire in the reactor building that forces contractor personnel who are doing a safety-related modification to leave, but the fire did not hamper operations personnel or equipment, would that fire be reportable?

Answer: No. The fire would not be reportable if the fire was not severe enough that it posed an actual threat to the plant and the delay in completing the modification did not significantly threaten the safe operation of the plant.

3.2.11 Transport of a Contaminated Person Offsite

§ 50.72(b)(3)(xii)	§ 50.73
"Any event requiring the transport of a radioactively contaminated person to an offsite medical facility for treatment."	There is no corresponding requirement in 10 CFR 50.73.

If not reported under 10 CFR 50.72(a), (b)(1), or (b)(2), an ENS notification is required under 10 CFR 50.72(b)(3) (an 8-hour report) for transport of a radioactively contaminated person to an offsite medical facility for treatment. The transport of a contaminated person offsite should be apparent at the time of occurrence. Therefore, if all events are reported properly, it is expected that all reports under 10 CFR 50.72 are as a result of an on-going condition.

Discussion

The phrase "radioactively contaminated" refers to either radioactively contaminated clothing or persons. If there is a potential for contamination (e.g., an initial onsite survey for radioactive contamination is required but has not been completed before transport of the person off site for medical treatment), the licensee should make an ENS notification. See the example.

---

[10] It is theoretically possible to have a control room fire that is discovered and extinguished quickly and, even in this location, does not significantly hamper the operators and does not threaten plant safety. Examples could include small paper fires in ashtrays or trash cans, or cigarette burns of furniture or upholstery.

No LER is required for transporting a radioactively contaminated person to an offsite medical facility for treatment.

Example

Radioactively Contaminated Person Transported Offsite for Medical Treatment

A contract worker experienced a back injury lifting a tool while working in a contaminated area and was considered potentially contaminated because his back could not be surveyed. Health physics technicians accompanied the worker to the hospital. The licensee made an ENS notification immediately and an update notification after clothing, but not the individual, was found to be contaminated. The HP technicians returned to the plant with the contaminated protective clothing worn by the worker.

An ENS notification is required because of the transport of a radioactively contaminated person to an offsite medical facility for treatment.

3.2.12 News Release or Notification of Other Government Agency

§ 50.72(b)(2)(xi)	§ 50.73
"Any event or situation, related to the health and safety of the public or on-site personnel, or protection of the environment, for which a news release is planned or notification to other government agencies has been or will be made. Such an event may include an on-site fatality or inadvertent release of radioactively contaminated materials."	There is no corresponding requirement in 10 CFR 50.73.

If not reported under 10 CFR 50.72(a) or (b)(1), licensees are required to notify the NRC via the ENS under 10 CFR 50.72(b)(2) (a 4-hour report). A news release or notification of other government agency should be apparent at the time of occurrence. Therefore, if all events are reported properly, it is expected that all reports under 10 CFR 50.72 are as a result of an on-going condition.

In the case of an event for which a news release is planned, the purpose of the report is to provide timely and accurate information so that the NRC can respond to heightened public concern. Accordingly, it is requested that the report be provided by the time the news release is issued.

Discussion

The purpose of this criterion is to ensure that the NRC is made aware of issues that will cause heightened public or government concern related to the radiological health and safety of the public or onsite personnel or protection of the environment.

Licensees typically issue press releases or notify local, county, State, or Federal agencies on a wide range of topics that are of interest to the general public. The NRC Operations Center does not need to be made aware of every press release made by a licensee. The following clarifications are intended to set a reporting threshold that ensures necessary reporting while minimizing unnecessary reporting.

Examples of events likely to be reportable under this criterion include the following:

- release of radioactively contaminated tools or equipment to public areas
- unusual or abnormal releases of radioactive effluents
- onsite fatality

Licensees generally do not have to report media and government interactions unless they are related to the radiological health and safety of the public or onsite personnel, or protection of the environment. For example, the NRC does not generally need to be informed under this criterion of the following:

- minor deviations from sewage or chlorine effluent limits
- minor nonradioactive, onsite chemical spills
- minor oil spills
- problems with plant stack or water tower aviation lighting
- peaceful demonstrations
- routine reports of effluent releases to other agencies
- releases of water from dams associated with the plant

For events or situations related to the health and safety of the public or onsite personnel, or to protection of the environment, licensees are required to notify the NRC within 4 hours of whichever of the following occurs first:

- a plan to report to either the press or another government agency is approved by an individual authorized to make the final decision

- a report has actually been made to the press or another government agency

Press Release

The NRC has an obligation to inform the public about issues within the NRC's purview that affect or raise a concern about the public health and safety. Thus, the NRC needs accurate, detailed information in a timely manner regarding such situations. The NRC should be aware of information that is available for the press or other government agencies.

However, the NRC need not be notified of every press release a licensee issues. The field of NRC interest is narrowed by the phrase "related to the health and safety of the public or onsite personnel, or protection of the environment" in order to exclude administrative matters or those events of no significance.

Routine radiation releases are not specifically reportable under this criterion. However, if a release receives media attention, the release is reportable under this criterion.

If possible, licensees should make an ENS notification before issuing a press release because news media representatives will usually contact the NRC public affairs officer shortly after its issuance for verification, explanation, or interpretation of the facts.

Other Government Notifications

For reporting purposes, "other government agencies" refers to local, State, or other Federal agencies.

Notifying another Federal agency does not relieve the licensee of the requirement to report to the NRC.

Some plants provide a State incident response facility with alarm indication coincident with control room alarms; e.g., an effluent radiation monitor alarm. However, an alarm received at a State facility is in itself not a requirement for notifying the NRC under this criterion. A release is reportable under this criterion if a press release is planned or a specific report (beyond the automatic alarm indication) has been or will be made to a State agency.

Examples

(1)   Onsite Drowning Government Notifications and Press Release

A man fell into the discharge canal while fishing and failed to resurface. The licensee notified the local sheriff, State Police, U.S. Coast Guard, and State emergency agencies. Local news agencies were granted onsite access for coverage of the event. The licensee notified the NRC resident inspector.

An ENS notification is needed because of the fatality on site, the other government notifications made, and media involvement.

(2)   Licensee Media Inquiries Regarding NRC Findings

As a result of a local newspaper article regarding the findings of an NRC regional inspection of the fire protection program under Appendix R, "Fire Protection Program for Nuclear Power Facilities Operating Prior to January 1, 1979," to 10 CFR Part 50, a licensee representative was interviewed on local television and radio stations. The licensee notified State officials and the NRC resident inspector.

The staff does not consider an ENS notification to be needed because the subject of the radio and TV interviews was an NRC inspection.

(3)   County Government Notification

The licensee informed county governments and other organizations of a spurious actuation of several emergency response sirens in a county (for about 5 minutes, according to county residents). The licensee also planned to issue a press release.

An ENS notification is needed because county agencies were notified about the inadvertent actuation of part of the public notification system. Such an event also would be reportable if the county informs the licensee of the problem because of the concern of members of the public for their radiological health and safety.

(4)   State Notification of Unscheduled Radiation Release

The licensee reported to the State that it was going to release about 50 curies of gaseous radioactivity to the atmosphere while filling and venting the pressurizer. The licensee then revised its estimate of the release to 153 curies. However, because the licensee had not informed the State within 24 hours of making the release, it had to reclassify the release as "unscheduled" per its agreement with the State. The licensee notified the State and the NRC resident inspector.

An ENS notification is needed because of the State notification of an "unscheduled" release of gaseous radioactivity. The initial notification to the State of the scheduled release does not need an ENS notification because it is considered to be a routine notification.

(5)   State Notification of Improper Dumping of Radioactive Waste

The licensee transported two secondary side filters to the city dump as nonradioactive waste but later determined that they were radioactive. The dump site was closed and the filters retrieved. The licensee notified the appropriate State agency and the NRC resident inspector.

An ENS notification is needed because of the notification to the State agency of the inadvertent release of radioactively contaminated material off site, which affects the radiological health and safety of the public and environment.

(6)   Reports Regarding Endangered Species

The licensee notified the U.S. Fish & Wildlife Service and a State agency that an endangered species of sea turtle was found in its circulating water structure trash bar. No press release was planned.

An ENS notification is required because of the notification of State and Federal agencies regarding the taking of an endangered species. (The NRC has statutory responsibilities regarding protection of endangered species.)

(7)   Routine Agency Notifications

A licensee notified the U.S. Environmental Protection Agency (EPA) that the circulation water temperature rise exceeded the release permit allowable. This event was caused by the unexpected loss of a circulating water pump while operating at 92-percent power. The licensee reduced power to 73 percent so that the circulating water temperature would decrease to within the allowable limits until the pump could be repaired.

A licensee notified the Federal Aviation Agency that it removed part of its auxiliary boiler stack aviation lighting from service to replace a faulty relay.

A licensee notified the State, EPA, U.S. Coast Guard, and U.S. Department of Transportation that 5 gallons of diesel fuel oil had spilled onto gravel-covered ground inside the protected area. The spill was cleaned up by removing the gravel and dirt.

The staff does not consider an ENS notification to be needed because these events are routine and have little significance.

3.2.13 Loss of Emergency Preparedness Capabilities

§ 50.72(b)(3)(xiii)	§ 50.73
"Any event that results in a major loss of emergency assessment capability, offsite response capability, or offsite communications capability (e.g., significant portion of control room indication, Emergency Notification System, or offsite notification system)."	There is no corresponding requirement in 10 CFR 50.73.

If not reported under 10 CFR 50.72(a), (b)(1), or (b)(2), an ENS notification is required under 10 CFR 50.72(b)(3) (an 8-hour report) for a major loss of emergency assessment, offsite response, or offsite communications capability. The loss of emergency preparedness capabilities should be apparent at the time of occurrence or discovery. Therefore, if all events are reported properly, it is expected that all reports under 10 CFR 50.72 are as a result of an on-going condition.

Discussion

This reporting requirement pertains to events that would result in a major loss of emergency assessment capability, offsite response capability, or offsite communications capabilities. The loss of these capabilities could substantially impair a licensee's, or offsite officials', ability to respond to an emergency if one were to occur or has occurred. The focus of this reporting requirement is in the loss of capabilities to perform functions identified in the respective emergency plan. Failures of individual systems or facilities that comprise these capabilities are reportable only to the extent that these failures meet the above threshold.

Notifying the NRC of these events permits the NRC to consider implementing compensatory measures and to more completely assess the consequences of such a loss should it occur during an accident or emergency. The following are examples of equipment or facilities that may be encompassed by this reporting requirement:

* Emergency Assessment Capabilities
    - safety parameter display system (SPDS)
    - primary emergency response facilities (ERFs)
    - plant monitors necessary for accident assessment

* Offsite Response Capabilities
    - public prompt notification system(s) including sirens (primary system)

* Offsite Communication Capabilities
    - ENS
    - other emergency communications facilities and equipment used between the licensee's onsite and offsite ERFs, and between the licensee and offsite officials

Losses of the above equipment and other situations should be evaluated for reportability as discussed below.

Loss of Emergency Assessment Capability

A major loss of emergency assessment capability includes those events that would significantly impair the licensee's emergency assessment capability if an emergency were to occur. Some engineering judgment is needed to determine the significance of the loss of particular equipment. For example, the loss of the SPDS alone may not need to be reported, but a loss of the SPDS concurrent with other plant indicators or annunciators being unavailable is reportable if the licensee would be unable to assess or monitor an accident or transient in progress. Examples of events that should be evaluated against this threshold for reportability include, but are not limited to:

- A loss of a significant portion of control room indication, including annunciators or monitors, or the loss of all plant vent stack radiation monitors, should be evaluated for reportability. In evaluating the reportability of such events, only those display systems, indicators, and annunciators that are relied upon in the emergency plan and the emergency plan implementing procedures addressing classification, assessment, or protective actions; or relied upon in other station procedures that provide input to these activities need to be considered. The indication remaining available should be considered in determining if a major loss of emergency assessment capability has occurred.

- A significant degradation in the licensee's ability to perform accident assessment functions assigned to a licensee primary ERF by the emergency plan. Typically, these functions would be the technical support center (TSC) but may include the emergency operations facility (EOF). Degradations would not be reportable if the ERF's assessment capabilities were restored to service within the facility activation times specified in the emergency plan. Planned maintenance which impacts the accident assessment functions of the ERF, or its supporting systems, need not be reported if (1) the ERF's assessment capabilities could be restored to service within the facility acitvation time specified in the emergency plan in the event of an accident or the licensee had implemented viable compensatory actions,[11] and (2) the planned outage is not expected to, and subsequently did not, exceed 72 hours.

Loss of Offsite Response Capability

A major loss of offsite response capability includes those events that would significantly impair the ability of the licensee or offsite officials to implement the functions of their respective emergency plans if an emergency were to occur. Examples of events that should be evaluated against this threshold for reportability include, but are not limited to:

- The occurrence of a significant natural hazard (e.g., earthquake, hurricane, tornado, flood, major winter storms) or other event that would do one or both of the following:

---

[11] "Promptly" means within the licensee's emergency plan requirements for facility activation time. A "viable" compensatory action is one that (1) can restore the required function in a reasonably comparable manner and (2) is proceduralized prior to an event.

- Prevent State and local jurisdictions from maintaining evacuation routes passable, or from maintaining other parts of the response infrastructure available, to the extent that these jurisdictions would be unable to implement the public protective measures called for in their emergency plan, if known by the licensee, or,

- Restrict access to the licensee's site, or its offsite primary EOFs, such that the licensee would not be able to augment its on-shift staff or activate its ERFs as required by the emergency plan. Offsite response support relied upon in the emergency plan, such as fire departments, local law enforcement, and ambulance services, would not be able to access the site.

Traffic impediments, such as fog, snow, and ice, should generally not be reported if they are within the respective capabilities of the licensee, State, or local officials to resolve or mitigate. Rather, the reporting requirement is intended to apply to more significant cases, such as the conditions around the Turkey Point plant after Hurricane Andrew struck in 1992 or the conditions around the Cooper station during the Midwest floods of 1993.

- Failures in the primary public alerting systems (e.g., sirens, tone alert radios), for whatever reason, that result in the loss of the capability to alert a large segment of the population in the emergency planning zone (EPZ) for more than 1 hour. The licensee should take reasonable measures to remain informed of the status of the primary public alerting system, regardless of who maintains the system, and must notify the NRC if the established thresholds are exceeded. A planned outage of the primary public alerting system need not be reported if (1) the licensee had arranged for the implementation of Federal Emergency Management Agency (FEMA)-approved backup alerting methods should public alerting become necessary, and (2) the planned outage is not expected to, and subsequently did not, exceed 24 hours.

Loss of Communications Capability

A major loss of communications capability include those events that would significantly impair the ability of the licensee to implement the functions of its emergency plans if an emergency were to occur. Failures of individual communications systems are reportable only to the extent that these failures meet this threshold. The failure of a single communication system need not be reported if there are viable alternative methods[12] of communicating information about the emergency.

This reporting requirement only addresses those communication systems that enable a licensee to make notifications and provide followup information to Federal, State, and local officials located offsite. It also includes communication capabilities between the site and licensee ERO personnel assigned offsite. Examples of communication systems whose failures should be evaluated against the above threshold for reportability include, but are not limited to, the following:

- ERDS

---

[12] A "viable" alternative method (or compensatory action) is one that (1) can perform the required function in a reasonably comparable manner and (2) is proceduralized prior to an event.

- ENS
- health physics network (HPN)
- other offsite communication systems, including the following:

    - dedicated telephone communication link to State or local officials

    - dedicated voice and data links between the site and emergency offsite response facilities

    - licensee radio system for communicating with offsite field monitoring teams

    - commercial telephone lines that are relied upon for use in emergency response

Each site's communications system will be different, and the significance of the loss of any one communication system may differ from site to site. This reporting requirement is intended to apply to serious conditions during which the telecommunications system can no longer fulfill the communications requirements of the emergency plan.

Planned maintenance that impacts the emergency communications capability need not be reported if (1) the communication system could be restored to service promptly in the event of an accident or the licensee had implemented viable compensatory actions,[13] and (2) the planned outage is not expected to, and subsequently did not, exceed 72 hours.

Although a notification may not be required under 10 CFR 50.72(b)(3(xiii) in the event of a loss of the ENS, HPN, or ERDS because of the availability of viable alternative communication means, the licensee should inform the NRC Operations Center of any failure of these systems so that the NRC may arrange for repair of NRC-supplied communications equipment. When informing the NRC Operations Center, licensees should use the commercial telephone number 301-816-5100. If the Operations Center (or the ERDS Data Center) notifies the licensee that an ENS, HPN, or ERDS line is out of service, there is no need for an additional call. At the time the failure is reported, the licensee should be prepared to supply the following information to expedite repair: (1) name of contact at location of failure, (2) commercial phone number of contact, (3) location of contact (i.e., street address, building number, room number, etc., and (4) any other information that would expedite repair.

Examples

(1)   Loss of Public Prompt Notification System

The NRC has not established a numerical threshold (e.g., number, percentage, or area of failed sirens) for this reporting requirement because the thresholds need to be specific to the particular EPZ. The NRC expects its licensees to establish thresholds that reflect the EPZ-specific population density and distribution, the locations of the sirens or other alerting devices, and the overlap in coverage of adjacent sirens. For example, a loss of 10 percent of the sirens in a high-density population area may have greater impact than

---

[13]   "Promptly" means within the emergency plan requirements specified for the communication system. A loss of the ability to make initial notifications would need to be restored within 15 minutes, while a loss of the ability to communicate between ERFs would need to be restored within the facility activation time. A "viable" compensatory action is one that (1) can restore the required function in a reasonably comparable manner, and (2) is proceduralized prior to the event.

50 percent of the sirens lost in a low-density area. Similarly, a loss of 10 percent of the sirens dispersed across the entire EPZ may not be as significant as losing the same number of sirens in a single jurisdiction. As such, notifications of the loss of the primary public prompt notification system will vary according to the licensee's "major loss" threshold. Previous notifications have included the following:

- 12 of 40 county alert sirens were disabled because of loss of power as a result of severe weather

- 28 of 54 alert sirens were reported out of service as a result of a local ice storm

- all offsite emergency sirens were:

  – found out of service during a monthly test
  – taken out of service for repair
  – out of service because control panel power was lost
  – out of service because the county radio transmitter failed

Failures in the primary public alerting systems (e.g., sirens, tone alert radios), for whatever reason, that result in the loss of the capability to alert a large segment of the population in the EPZ for more than 1 hour should be reported as a major loss of offsite response capability. However, a planned outage need not be reported if (1) the licensee had arranged for the implementation of FEMA-approved backup alerting methods should public alerting become necessary, and (2) the planned outage is not expected to, and subsequently did not, exceed 24 hours. No LER is required because there are no corresponding 10 CFR 50.73 requirements.

(2)   Loss of ENS and Commercial Telephone System

The licensee determined that ENS and commercial telecommunications capability was lost to the control room when a fiber optic cable was severed during maintenance. A communications link was established and maintained between the site and the load dispatcher via microwave transmission. Both the ENS and commercial communications capability were restored approximately 90 minutes later.

An ENS notification is required because of the major loss of communications capability. Although the microwave link to the site was established and maintained during the telephone outage, the link would not allow direct communication between the NRC and the control room, and therefore, this in itself does not fully compensate for the loss of communication that would be required in the event of an emergency at the plant. No LER is required because there are no corresponding 10 CFR 50.73 requirements.

(3)   Loss of Direct Communication Line to Police

The licensee determined that the direct telephone line to the State Police had been out of service. In this example, no ENS notification is required because commercial telephone lines to the State Police were available. An ENS notification would be required if the loss of the direct telephone line(s) to various police, local, or State emergency or regulatory agencies is not compensated for by other readily available offsite communications systems. No LER is required because there are no corresponding 10 CFR 50.73 requirements.

(4)   <u>Loss of the Emergency Response Data System</u>

The licensee determined that the ERDS was out of service due to a failure of licensee-owned and -maintained equipment. However, the ENS was available. Because the ERDS is identified as a supplement to the ENS in Appendix E of 10 CFR 50, the failure of the ERDS does not constitute a major loss of offsite communication capability provided that the ENS is available and, as a result, no report under this reporting criterion is required. If, however, the failure is determined to be in NRC-maintained equipment, the licensee should inform the ERDS help desk of the outage so that the NRC can arrange for repair.

3.2.14 Single Cause that Could Have Prevented Fulfillment of the Safety Functions of Trains or Channels in Different Systems

§ 50.72	§ 50.73(a)(2)(ix)
There is no corresponding requirement in 10 CFR 50.72.	"(A) Any event or condition that as a result of a single cause could have prevented the fulfillment of a safety function for two or more trains or channels in different systems that are needed to: (1) Shut down the reactor and maintain it in a safe shutdown condition; (2) Remove residual heat; (3) Control the release of radioactive material; or (4) Mitigate the consequences of an accident. (B) Events covered in paragraph (ix)(A) of this section may include cases of procedural error, equipment failure, and/or discovery of a design, analysis, fabrication, construction, and/or procedural inadequacy. However, licensees are not required to report an event pursuant to paragraph (ix)(A) of this section if the event results from: (1) A shared dependency among trains or channels that is a natural or expected consequence of the approved plant design; or (2) Normal and expected wear or degradation."

An LER is required for an event that meets the conditions stated in the rule.

Discussion

The level of judgment for reporting an event or condition under this criterion is a reasonable expectation of preventing fulfillment of a safety function. In the discussions that follow, several different expressions, such as "would have," "could have," "alone could have," and "reasonable doubt," are used to characterize this standard. In the staff's view, all of these should be judged on the basis of a reasonable expectation of preventing fulfillment of the safety function. For trains or channels that have been declared inoperable, the capability is considered degraded to a point where it cannot perform with reasonable expectation or reliability. As a result, subject to the exceptions stated in 10 CFR 50.73(a)(2)(ix)(B)(1) and (2), for trains or channels within the scope of this criterion, a report is required when there is a determination that two or more trains or channels in different systems are inoperable as a result of a single cause while in a required mode or other specified condition in the TS Applicability. However, reports are not required when trains or channels in different systems are declared inoperable as part of a planned evolution for maintenance or surveillance testing when done in accordance with an approved procedure and the plant's TS (unless a condition is discovered that would have resulted in the trains or channels in different systems being declared inoperable as a result of a single cause). For guidance on determining whether a train or channel is operable, see RIS 2005-20, Revision 1. Operable but nonconforming or degraded conditions are not considered reportable under this criterion.

The intent of this criterion is to capture those events in which, as a result of a single cause, there would have been a failure of two or more trains or channels to properly complete their safety function, regardless of whether there was an actual demand. For example, if, as a result of a single cause, a train of the HPSI system and a train of the AFW system failed, the event would be reportable even if there was no demand for the systems' safety functions.

Examples of a single cause responsible for a reportable event may include cases of procedural error, equipment failure, or discovery of a design, analysis, fabrication, construction, or procedural inadequacy. They may also include such factors as high ambient temperatures, heatup from energization, inadequate preventive maintenance, oil contamination of air systems, incorrect lubrication, or use of nonqualified components.

The event is reportable if, as a result of a single cause, two or more trains or channels are inoperable, regardless of whether the problem was discovered in both trains at the same time.

Trains or channels, for reportability purposes, are defined as those trains or channels designed to provide protection against single failures. Many systems containing active components are designed as at least a two-train system. Each train in a two-train system can normally satisfy all of the system functions.

SSCs within scope include only safety-related SSCs required by the TS to be operable that are intended to mitigate the consequences of an accident as discussed in Chapters 6 and 15 of the FSAR (or equivalent chapters). Accidents are identified as events of moderate frequency, infrequent incidents, or limiting faults as discussed in Regulatory Guide 1.70 (or equivalent classifications of the three types of events). ANS categorizes these events as Condition II, III and IV type events.

Examples

(1)   Solenoid-Operated Valve Deficiency

During testing, two containment isolation valves failed to function as a result of improper air gaps in the solenoid-operated valves that controlled the supply of instrument air to the containment isolation valves.

The valves were powered from the same electrical division. Therefore, 10 CFR 50.73(a)(2)(vii) (common cause inoperability of independent trains or channels) would not apply. The two valves isolated fluid process lines in two different systems. Thus, 10 CFR 50.73(a)(2)(v) (condition that could have prevented fulfillment of the safety function of a structure or system) would apply only if engineering judgment indicates that there was a reasonable expectation of preventing fulfillment of the safety function for redundant valves within the same system.[14] However, this criterion would certainly apply if a single cause (such as a design inadequacy) induced the improper air gaps, thus preventing fulfillment of the safety function of two trains or channels in different systems.

(2)   Degraded Valve Stems

A motor-operated valve in one train of a system was found with a crack 75 percent through the stem. Although the valve stem did not fail, engineering evaluation indicated that further cracking would occur that could have prevented fulfillment of its safety function. As a result, the train was not considered capable of performing its specified safety function. The valve stem was replaced with a new one.

The root cause was determined to be environmentally assisted stress-corrosion cracking that resulted from installation of an inadequate material some years earlier. The same inadequate material had been installed in a similar valve in a different system at the same time. The similar valve was exposed to similar environmental conditions as the first valve.

The condition is reportable under this criterion if engineering judgment indicates that there was a reasonable expectation of preventing fulfillment of the safety function of both affected trains. This depends on details such as whether the second valve stem was also significantly degraded and, if not, whether any future degradation of the second valve stem would have been discovered and corrected, as a result of routine maintenance programs, before it could become problematic.

(3)   Overpressure Due to Thermal Expansion

It was determined that a number of liquid-filled and isolated containment penetration lines in multiple safety systems were not adequately designed to accommodate the internal pressure buildup that could occur because of thermal expansion caused by heatup after a design-basis accident. The problem existed because the original design failed to consider this effect following a postulated accident.

---

[14]   Or, alternatively, that there was reasonable doubt that the safety function would have been fulfilled if the affected trains had been called upon to perform them.

The condition is reportable under this criterion because there was a reasonable expectation of preventing fulfillment of the safety function of multiple trains or channels as a result of a single cause.

(4)   Cable Degradation

One of three component cooling water pumps tripped due to a ground fault on a power cable leading to the pump. The likely cause was determined to be moisture permeation into the cable insulation over time in a section of cable that was exposed to water.

The event is reportable under this criterion if engineering judgment indicates that there was a reasonable expectation of preventing fulfillment of the safety function of an additional train in a different system as a result of the same cause. For example, if cable testing indicates that another cable to safety-related equipment was likely to fail as a result of the same cause, the event is reportable.

(5)   Overstressed Valve Yokes

It was determined that numerous motor-operated valve yokes experienced overthrusting that exceeded design-basis stress levels. The cause was lack of knowledge that resulted in inadequate design engineering at the time the designs were performed.

Some of the motor-operated valve yokes, in different systems, were being overstressed enough during routine operations that, although they were currently capable of performing their specified safety functions, the overstressing would, with the passage of time, render them incapable of performing those functions. The condition is reportable under this criterion if engineering judgment indicates that there was a reasonable expectation of preventing fulfillment of the safety function of trains or channels in two or more different systems.[15]

(6)   Heat Exchanger Fouling

Periodic monitoring of heat exchanger performance indicated that two heat exchangers in two different systems required cleaning in order to ensure they would remain operable. The degree of fouling was within the range of the normal expectations upon which the monitoring and maintenance procedures were based.

The event is not reportable under this criterion because there was not a reasonable expectation of preventing the fulfillment of the safety function of the heat exchangers.

(7)   Pump Vibration

Based on increasing vibration trends, identified by routine vibration monitoring, it was determined that a pump's bearings required replacement. Other pumps in different systems with similar designs and service histories experience similar bearing degradation. However, it is expected that the degradation will be detected and corrected before failure occurs.

---

[15]   Or, alternatively, there was reasonable doubt that the safety function would have been fulfilled if the affected trains had been called upon to perform them.

Such bearing degradation is not reportable under this criterion because it is normal and expected.

3.3 Followup Notification

This section addresses 10 CFR 50.72(c), "Followup Notification." These notifications are in addition to making the required initial telephone notifications under 10 CFR 50.72(a) or (b). Reporting under this paragraph is intended to provide the NRC with timely notification when an event becomes more serious or additional information or new analysis clarifies an event. The paragraph also authorizes the NRC to maintain a continuous communications channel for acquiring necessary followup information.

§ 50.72(c)	§ 50.73
"*Followup notification.* With respect to the telephone notifications made under paragraphs (a) and (b) of this section, in addition to making the required initial notification, each licensee shall, during the course of the event: (1) *Immediately report* (i) any further degradation in the level of safety of the plant or other worsening plant conditions, including those that require the declaration of any of the Emergency Classes, if such a declaration has not been previously made, or (ii) any change from one Emergency Class to another, or (iii) a termination of the Emergency Class. (2) *Immediately report* (i) the results of ensuing evaluations or assessments of plant conditions, (ii) the effectiveness of response or protective measures taken, and (iii) information related to plant behavior that is not understood. (3) Maintain an open, continuous communication channel with the NRC Operations Center upon request by the NRC."	There is no corresponding requirement in 10 CFR 50.73.

Discussion

These criteria are intended to provide the NRC with timely notification when an event becomes more serious or additional information or new analyses clarify an event. They also permit the NRC to maintain a continuous communications channel because of the need for continuing followup information or because of telecommunications problems.

With regard to the open, continuous communications channel, licensees have a responsibility to provide enough on-shift personnel, knowledgeable about plant operations and emergency plan implementation, to enable timely, accurate, and reliable reporting of operating events without interfering with plant operation as discussed in the Statements of Consideration for the rule (48 FR 33850 published on July 26, 1983, 48 FR 39039 published on August 29, 1983, and 65 FR 63769 published on October 25, 2000) and IN 85-80, "Timely Declaration of an Emergency Class, Implementation of an Emergency Plan, and Emergency Notifications," dated October 15, 1985.

4. EMERGENCY NOTIFICATION SYSTEM REPORTING

This section describes the ENS referenced in 10 CFR 50.72 and provides general and specific guidelines for ENS reporting.

4.1 Emergency Notification System

The NRC Operations Center is the nucleus of the ENS and has the capability to handle emergency communication needs. The NRC's response to both emergencies and nonemergencies is coordinated in this communication center. The key NRC emergency communications personnel, the emergency officer (EO), regional duty officer (RDO), and the headquarters operations officer (HOO), are trained to notify appropriate NRC personnel and to focus appropriate NRC management attention on any significant event.

(1)   Emergency Notification System Telephones

Each commercial nuclear power reactor facility has ENS telephones. These telephones are located in each licensee's control room, TSC, and EOF. A separate ENS line is installed at EOFs that are not on site.

(2)   Health Physics Network Telephones

The HPN is designed to provide health physics and environmental information to the NRC Operations Center in the event of an ongoing emergency. These telephones are installed in each licensee's TSC and EOF.

(3)   Recording

The NRC records all conversations with the NRC Operations Center. The recordings are saved for 1 month in case there is a public or private inquiry.

(4)   Facsimile Transmission (Fax)

Licensees occasionally fax an event notification into the NRC Operations Center on a commercial telephone line in conjunction with making an ENS notification. However, 10 CFR 50.72 requires that licensees notify the NRC Operations Center via the ENS; therefore, licensees also must make an ENS notification.

4.2 General ENS Notification

4.2.1 Timeliness

The required timing for ENS reporting is spelled out in 10 CFR 50.72(a)(3), (b)(1), (b)(2), (b)(3), (c)(1), and (c)(2) as "immediate" and "as soon as practical and in all cases within one (or four or eight) hour(s)" of the occurrence of an event (depending on its significance and the need for prompt NRC action). The intent is to require licensees to make and act on reportability decisions in a timely manner so that ENS notifications are made to the NRC as soon as practical, keeping in mind the safety of the plant. See Section 2.5 for further discussion of reporting timeliness.

4.2.2 Voluntary Notifications

Licensees may make voluntary or courtesy ENS notifications about events or conditions in which the NRC may be interested. The NRC responds to any voluntary notification of an event or condition as its safety significance warrants, regardless of the licensee's classification of the reporting requirement. If it is determined later that the event is reportable, the licensee can change the ENS notification to a required notification under the appropriate 10 CFR 50.72 reporting criterion.

4.2.3 ENS Notification Retraction

If a licensee makes a 10 CFR 50.72 ENS notification and later determines that the event or condition was not reportable, the licensee should call the NRC Operations Center on the ENS telephone to retract the notification and explain the rationale for that decision. There is no set time limit for ENS telephone retractions. However, because most retractions occur following completion of engineering and/or management review, it is expected that retractions would occur shortly after such review. A retracted ENS report is retained in the ENS database, along with the retraction. See Section 2.8 for further discussion of retractions.

4.2.4 ENS Event Notification Worksheet (NRC Form 361)

The ENS "Event Notification Worksheet" (NRC Form 361) provides the usual order of questions and discussion for easier communication and its use often enables a licensee to prepare answers for a more clear and complete notification. A clear ENS notification helps the HOO to understand the safety significance of the event. Licensees may obtain an event number and notification time from the HOO when the ENS notification is made. If an LER is required, the licensee may include this information in the LER to provide a cross-reference to the ENS notification, making the event easier to trace.

Licensees should use proper names for systems and components, as well as their alphanumeric identifications, during ENS notifications. Licensees should avoid using local jargon for plant components, areas, operations, and the like so that the HOO can quickly understand the situation and have fewer questions. In addition, others not familiar with the plant can more readily understand the situation.

Electronic versions of NRC Form 361 are available at the NRC public Web site at http://www.nrc.gov/reading-rm/doc-collections/forms/.

4.3 Typical ENS Reporting Issues

At the time of an ENS notification, the NRC must independently assess the status of the reactor to determine if it is in a safe condition and expected to remain so. The HOO needs to understand the safety significance of each event to brief NRC management or initiate an NRC response. The HOO will be primarily concerned about the safety significance of the event, the current condition of the plant, and the possible near-term effects the event could have on plant safety. The HOO will attempt to obtain as complete a description as is available at the time of the notification of the event or condition, its causes, and its effects. Depending on the licensee's description of the event, the HOO may be concerned about other related issues. The questions that licensees typically may be asked to discuss do not represent a requirement for reporting. These questions are of a nature to allow the HOO information to more fully understand the event and its safety significance and are not meant in any way to distract the licensee from more important issues.

The licensee's first responsibility during a transient is to stabilize the plant and keep it safe. However, licensees should not delay in declaring an emergency class when conditions warrant because delaying the declaration can defeat the appropriate response to an emergency. Because of the safety significance of a declared emergency, time is of the essence. The NRC needs to become aware of the situation as soon as practical to activate the NRC Operations Center and the appropriate NRC regional incident response center, as necessary, and to notify other Federal agencies.

The effectiveness of the NRC response during an event depends largely on complete and accurate reporting from the licensee. During an emergency, the appropriate regional incident response center and the NRC Operations Center become focal points for NRC action. Licensee actions during an emergency are monitored by the NRC to ensure that appropriate action is being taken to protect the health and safety of the public. When required, the NRC supports the licensee with technical analysis and coordinates logistics support. The NRC keeps other Federal agencies informed of the status of an incident and provides information to the media. In addition, the NRC assesses and, if necessary, confirms the appropriateness of actions recommended by the licensee to local and State authorities.

IN 85-80 indicates that it is the licensee's responsibility to ensure that adequate personnel, knowledgeable about plant conditions and emergency plan implementing procedures, are available on shift to assist the shift supervisor to classify an emergency and activate the emergency plan, including making appropriate notifications, without interfering with plant operation. When 10 CFR 50.72 was published, the NRC made clear its intent in the Statements of Consideration that notifications on the ENS to the NRC Operations Center should be made by those knowledgeable of the event. If the description of any emergency is to be sufficiently accurate and timely to meet the intent of the NRC's regulations, the personnel responsible for notification must be properly trained and sufficiently knowledgeable of the event to report it correctly. The NRC did not intend that notifications made under 10 CFR 50.72 would be made by those who did not understand the event that they are reporting.

ENS reportability evaluations should be concluded and the ENS notification made as soon as practical and in all cases within 1, 4, or 8 hours to meet 10 CFR 50.72. The Statements of Consideration noted that the 1-hour deadline is necessary if the NRC is to fulfill its responsibilities during and following the most serious events occurring at operating nuclear power plants without interfering with the operator's ability to deal with an accident or transient in the first few critical minutes (48 FR 39041; August 29, 1983).

5. LICENSEE EVENT REPORTS

This section discusses the guidelines for preparing and submitting LERs. Section 5.1 addresses administrative requirements and provides guidelines for submittal; Section 5.2 addresses the requirements and guidelines for the LER content. Portions of the rule are quoted, followed by explanation, if necessary. Copies of the required "Licensee Event Report" form (NRC Form 366), "Licensee Event Report (LER) Continuation Sheet" (NRC Form 366A), and "Licensee Event Report (LER) Failure Continuation" form (NRC Form 366B) may be found at the NRC public Web site at http://www.nrc.gov/reading-rm/doc-collections/forms/.

5.1 LER Reporting Guidelines

This section addresses administrative requirements and provides guidelines for submittal. Topics addressed include submission of reports, forwarding letters, cancellation of LERs, report legibility, reports other than LERs that use LER forms, supplemental information, revised reports, and general instructions for completing LER forms.

5.1.1 Submission of LERs

---

§ 50.73(d)

"Submission of reports. Licensee Event Reports must be prepared on Form NRC 366 and submitted to the U.S. Nuclear Regulatory Commission, as specified in § 50.4."

An LER is to be submitted (mailed) within 60 days of the discovery date. If a 60-day period ends on a Saturday, Sunday, or holiday, reports submitted on the first working day following the end of the 60 days are acceptable. If a licensee knows that a report will be late or needs an additional day or so to complete the report, the situation should be discussed with the appropriate NRC regional office. See Section 2.5 for further discussion of discovery date.

5.1.2 LER Forwarding Letter and Cancellations

The cover letter forwarding an LER to the NRC should be signed by a responsible official. There is no prescribed format for the letter. The date the letter is issued and the report date should be the same. Licensees are encouraged to include the NRC resident inspector and the Institute of Nuclear Power Operations in their distribution. Multiple LERs can be forwarded by one forwarding letter.

Cancellations of LERs submitted should be made by letter. The letter should state that the LER is being canceled (i.e., formally withdrawn). The bases for the cancellation should be explained so that the staff can understand and review the reasons supporting the determination. The notice of cancellation will be filed and stored with the LER and acknowledgement made in various automated data systems. The LER will be removed from the LER database.

5.1.3 Report Legibility

> **§ 50.73(e)**
>
> "The reports and copies that licensees are required to submit to the Commission under the provisions of this section must be of sufficient quality to permit legible reproduction and micrographic processing."

No further explanation is necessary.

5.1.4 Voluntary LERs

Indicate information-type LERS (i.e., voluntary LERs) by checking the "Other" block in Item 11 of the LER form and type "Voluntary Report" in the space immediately below the block. Also give a sequential LER number to the voluntary report as noted in Section 5.2.7(6). Because not all requirements of 10 CFR 50.73(b), "Contents," may pertain to some voluntary reports, licensees should develop the content of such reports to best present the information associated with the situation being reported.

See Section 2.7 for additional discussion of voluntary LERs.

5.1.5 Supplemental Information and Revised LERs

> **§ 50.73(c)**
>
> "The Commission may require the licensee to submit specific additional information beyond that required by paragraph (b) of this section if the Commission finds that supplemental material is necessary for complete understanding of any unusually complex or significant event. These requests for supplemental information will be made in writing and the licensee shall submit, as specified in § 50.4, the requested information as a supplement to the initial LER."

This provision authorizes the NRC staff to require the licensee to submit specific supplemental information.

If an LER is incomplete at the time of original submittal or if it contains significant incorrect information of a technical nature, the licensee should use a revised report to provide the additional information or to correct technical errors discovered in the LER. Identify the revision to the original LER in the LER number as described in Section 5.2.7, Item (6).

The revision should be complete and should not contain only supplementary or revised information to the previous LER, because the revised LER will replace the previous report in the computer file. In addition, indicate in the text on the LER form the revised or supplementary information by placing a vertical line in the margin. If an LER mentions that an engineering study was being conducted, report the results of the study in a revised LER only if it would significantly change the reader's perception of the course, significance, implications, or consequences of the event or if it results in substantial changes in the corrective action planned by the licensee.

Use revisions only to provide additional or corrected information about a reported event. Do not use a revision to report subsequent failures of the same or like component, except as permitted in 10 CFR 50.73. Some licensees have incorrectly used revisions to report new events that were discovered months after the original event because they were loosely related to the original event. These revisions had different event dates and discussed new, although similar, events. Report events of this type as new LERs and not as revisions to previous LERs.

5.1.6 Special Reports

There are a number of requirements in various sections of the TS that require reporting of operating experience that is not covered by 10 CFR 50.73. If LER forms are used to submit special reports, check the "Other" block in Item 11 of the form and type "Special Report" in the space immediately below the block. The provisions of 10 CFR 50.73(b) may not be applicable or appropriate in a special report. Develop the content of the report to best present the information associated with the situation being reported. In addition, if the LER form is used to submit a special report, use a report number from the sequence used for LERs.

If an event is reportable both under 10 CFR 50.73 and as a special report, check the block in Item 11 for the applicable section of 10 CFR 50.73 as well as the "Other" block for a special report. The content of the report should depend on the reportable situation.

5.1.7 Appendix J Reports (Containment Leak Rate Test Reports)

A licensee must perform containment integrated and local leak rate testing and report the results as required by Appendix J to 10 CFR Part 50. When the leak rate test identifies a situation reportable under 10 CFR 50.73 (see Section 3.2.4 of this report), submit an LER and include the results in a report under Appendix J to 10 CFR Part 50 by reference, if desired. The LER should address only the reportable situation, not the entire leak rate test.

5.1.8 10 CFR 73.71 Reports

Submit events or conditions that are reportable under 10 CFR 73.71, "Maintenance of Records, Making of Reports," using the LER forms with the appropriate blocks in Item 11 checked. If the report contains Safeguards Information as defined in 10 CFR 73.21, "Protection of Safeguards Information: Performance Requirements," the LER forms may still be used but should be appropriately marked in accordance with 10 CFR 73.21. Include safeguards and security information only in the narrative and not in the abstract. In addition, the text should clearly indicate the information that is safeguards or security information. Finally, the requirements of 10 CFR 73.21(g) must be met when transmitting Safeguards Information. For additional guidelines on 10 CFR 73.71 reporting, see Regulatory Guide 5.62, Revision 1, "Reporting of Safeguards Events," issued November 1987; NUREG-1304, "Reporting of Safeguards Events," issued February 1988; and Generic Letter 91-03, "Reporting of Safeguards Events," dated March 6, 1991.

If the LER contains proprietary information, mark it appropriately in Item 17 (text) on the LER form. Include proprietary information only in the narrative and not in the abstract. In addition, indicate clearly in the narrative the information that is proprietary. Finally, the requirements of 10 CFR 2.790(b) must be met when transmitting proprietary information.

5.1.9 Availability of LER Forms

The NRC will provide LER forms (i.e., NRC Forms 366, 366A, and 366B) free of charge. Copies may be obtained by writing to the NRC Records Management Branch, Office of the Chief Information Officer, U.S. Nuclear Regulatory Commission, Washington, DC 20555.

Electronic versions are also available at the NRC public Web site at http://www.nrc.gov/reading-rm/doc-collections/forms/.

5.2 LER Content Requirements and Preparation Guidance

5.2.1 Optical Character Reader

To help reduce the number of errors incurred by the optical character reader used to read LER contents into NRC databases, the NRC suggests the following practices.

The staff suggests that you do not use underscore, do not use bold print, do not use italic print style, do not end any lines with a hyphen, and do not use paragraph indents. Instead, print copy single spaced, with a blank line between paragraphs.

The following are limitations on the use of symbols in the textual areas:

- Spell out the word "degree."
- Use </= for "less than or equal to."
- Use >/= for "greater than or equal to."
- Use +/- for "plus or minus."
- Spell out all Greek letters.

Do not use exponents. A number should be either expressed as a decimal, spelled out, or, preferably, designated in terms of "E" (E field format). For example, 4.2×10^{-6} could be expressed as 4.2E-6, 0.0000042 or 4.2 x 10(-6).

Define all abbreviations and acronyms in both the text and the abstract and explain all component designators the first time they are used (e.g., the ESW pump 1-SW-P-1A).

5.2.2 Narrative Description or Text (NRC Form 366A, Item 17)

(1) General

> ### § 50.73(b)(2)(i)
>
> The LER shall contain the following: "A clear, specific, narrative description of what occurred so that knowledgeable readers conversant with the design of commercial nuclear power plants, but not familiar with the details of a particular plant, can understand the complete event."

There is no prescribed format for the LER text; write the narrative in a format that most clearly describes the event. After the narrative is written, however, review the appropriate sections of 10 CFR 50.73(b) to make sure that applicable subjects have been adequately addressed. It is helpful to use headings to improve readability. For example, some LERs employ major headings such as event description, safety consequences, corrective actions, and previous similar events and subheadings such as initial conditions, dates and times, event classification, systems status, event or condition causes, failure modes, method of discovery, component information, immediate corrective actions, and actions to prevent recurrence.

Explain exactly what happened during the entire event or condition, including how systems, components, and operating personnel performed. Do not cover specific hardware problems in excessive detail. Describe unique characteristics of a plant as well as other characteristics that influenced the event (favorably or unfavorably). Avoid using plant-unique terms and abbreviations, or, as a minimum, clearly define them. The audience for LERs is large and does not necessarily know the details of each plant.

Include the root causes, the plant status before the event, and the sequence of occurrences. Describe the event from the perspective of the operator (i.e., what the operator saw, did, perceived, understood, or misunderstood). Specific information that should be included, as appropriate, is described in 10 CFR 50.73(b)(2)(ii), (b)(3), (b)(4), and (b)(5) and separately in the following sections.

If several systems actuate during an event, describe all aspects of the complete event, including all actuations sequentially, and those aspects that by themselves would not be reportable. For example, if a single component failure (generally not reportable) occurs following a reactor scram (reportable), describe the component failure in the narrative of the LER for the reactor scram. It is necessary to discuss the performance and status of equipment important for defining and understanding what happened and for determining the potential implications of the event.

Paraphrase pertinent sections of the latest submitted FSAR rather than referencing them because not all organizations or individuals have access to FSARs. Extensive cross-referencing would be excessively time consuming, considering the large number of LERs and large number of reviewers that read each LER. Ensure that each applicable component's safety-significant effect on the event or condition is clearly and completely described.

Do not use statements such as "this event is not significant with respect to the health and safety of the public" without explaining the basis for the conclusion.

---

§ 50.73(b)(2)(ii)(A)

The narrative description must include the following: "Plant operating conditions before the event."

Describe the plant operating conditions such as power level or, if not at power, describe the mode, temperature, and pressure that existed before the event.

§ 50.73(b)(2)(ii)(B)

The narrative description must include the following: "Status of structures, components, or systems that were inoperable at the start of the event and that contributed to the event."

If there were no SSCs that were inoperable at the start of the event and contributed to the event, so state. Otherwise, identify SSCs that were inoperable and contributed to the initiation or limited the mitigation of the event. This should include alternative mitigating SSCs that are a part of normal or emergency operating procedures that were or could have been used to mitigate, reduce the consequences of, or limit the safety implications of the event. Include the impact of support systems on mitigating systems that could have been used.

§ 50.73(b)(2)(ii)(C)

The narrative description must include the following: "Dates and approximate times of occurrences."

For a transient or system actuation event, the event date and time are the date and time that the event actually occurred. If the event is a discovered condition for which the occurrence date is not known, the event date should be specified as the discovery date. However, a discussion of the best estimate of the event date and its basis should be provided in the narrative. For example, if a design deficiency was identified on March 27, 1997, that involved a component installed during refueling in the spring of 1986, and only the discovery date is known with certainty, the event date should be specified as the discovery date. A discussion should be provided that describes, based on the best information available, the most likely time that the design flaw was introduced into the component (e.g., by the manufacturer or by plant engineering prior to procurement). The length of time that the component was in service should also be provided (i.e., when it was installed).

Discuss both the discovery date and the event date if they differ. If an LER is not submitted within 60 days from the event date, explain the relationship between the event date, discovery date, and report date in the narrative. See Section 2.5 for further discussion of discovery date.

Give dates and approximate times for all major occurrences discussed in the LER (e.g., discoveries; immediate corrective actions; systems, components, or trains declared inoperable or operable; reactor trip; actuation and termination of equipment operation; and stable conditions achieved). In particular, for standby pumps and emergency generators, indicate the length of time of operation and any intermittent periods of shutdown or inoperability during the event. Include an estimate of the time and date of failure of systems, components, or trains if different from the time and date of discovery. A chronology may be used to clarify the timing of personnel and equipment actions.

For equipment that was inoperable at the start of the event, provide an estimate of the time the equipment became inoperable and the last time the equipment was demonstrated to be capable of performing its safety function. The licensee should provide the basis for this conclusion (e.g., a test was successfully run or the equipment was operating). For equipment that failed, provide the failure time and the last time the equipment was demonstrated to be capable of performing its safety function. The licensee should provide the basis for this conclusion (e.g., a test was performed or the equipment was operating).

Components such as valves and snubbers may be tested over a period of several weeks. During this period, a number of inoperable similar components may be discovered.[16] In such cases, similar failures that are reportable and that are discovered during a single test program within the 60 days of discovery of the first failure may be reported as one LER. For similar failures that are reportable under 10 CFR 50.73 criteria and that are discovered during a single test program or activity, report all failures that occurred within the first 60 days of discovery of the first failure on one LER. However, the 60-day clock starts when the first reportable event is discovered. State in the LER text (and code the information in Items 14 and 15) that a supplement to the LER will be submitted when the test is completed. Submit a revision to the original LER when the test is completed. Include all of the failures, including those reported in the original LER, in the revised LER (i.e., the revised LER should stand alone).

(2) Failures and Errors

---

§ 50.73(b)(2)(ii)(D)

The narrative description must include the following: "The cause of each component or system failure or personnel error, if known."

Include the root cause(s) identified for each component or system failure (or fault) or personnel error. Contributing factors may be discussed as appropriate. For example, a valve stem breaking could have been caused by a limit switch that had been improperly adjusted during maintenance; in this case, the root cause might be determined to be personnel error and additional discussion could focus on the limit switch adjustment. If the personnel error is determined to have been caused by deficient procedures or inadequate personnel training, this should be explained.

If the cause of a failure cannot be readily determined and the investigation is continuing, the licensee should indicate what additional investigation is planned. A supplemental LER should be submitted following the additional investigation if substantial information is identified that would significantly change a reader's perception of the course or consequences of the event, or if there are substantial changes in the corrective actions planned by the licensee.

---

§ 50.73(b)(2)(ii)(E)

The narrative description must include the following: "The failure mode, mechanism, and effect of each failed component, if known."

Include the failure mode, mechanism (immediate cause), and effect of each failed component in the narrative. The effect of the failure on safety systems and functions should be fully described. Identify the specific part that failed and the specific trains and systems rendered

---

[16] Note that inoperable similar components might indicate common cause failures of independent trains or channels, which are reportable under 10 CFR 50.73(a)(2)(vii); see Section 3.2.8 for further discussion.

inoperable or degraded. Identify all dependent systems rendered inoperable or degraded. Indicate whether redundant trains were operable and available.

If the equipment is degraded but not failed, the licensee should describe the degradation and its effects and indicate the basis for the conclusion that the equipment would still perform its intended function.

---

§ 50.73(b)(2)(ii)(F)

The narrative description must include the following: "The Energy Industry Identification System component function identifier and system name of each component or system referred to in the LER.

(1) The Energy Industry Identification System is defined in: IEEE Std 803-1983 (May 16, 1983) Recommended Practice for Unique Identification in Power Plants and Related Facilities--Principles and Definitions.

(2) IEEE Std 803-1983 has been approved for incorporation by reference by the Director of the Federal Register in accordance with 5 U.S.C. 552(a) and 1 CFR part 51.

(3) A notice of any changes made to the material incorporated by reference will be published in the Federal Register. Copies may be obtained from the Institute of Electrical and Electronics Engineers, 445 Hoes Lane, P.O. Box 1331, Piscataway, NJ 08855-1331. IEEE Std 803-1983 is available for inspection at the NRC's Technical Library, which is located in the Two White Flint North Building, 11545 Rockville Pike, Rockville, Maryland 20852-2738; or at the National Archives and Records Administration (NARA). For information on the availability of this material at NARA, call 202-741-6030, or go to: http://www.archives.gov/federal_register/code_of_federal_regulations/ibr_locations.html.

The system name may be either the full name (e.g., "reactor coolant system") or the two-letter system code (such as "AB" for the RCS). However, when the name is long (e.g., low-pressure coolant injection system), the system code (e.g., BO) should be used. If the full names are used, the Energy Industry Identification System (EIIS) component function identifier and/or system identifier (i.e., the two-letter code) should be included in parentheses following the first reference to a component or system in the narrative. The component function identifiers and system identifiers need not be repeated with each subsequent reference to the same component or system.

If a component within the scope of the Equipment Performance and Information Exchange (EPIX) System is involved, the system and train designation should be consistent with the EIIS used in EPIX.

---

§ 50.73(b)(2)(ii)(G)

The narrative description must include the following specific information as appropriate for the particular event: "For failures of components with multiple functions, include a list of systems or secondary functions that were also affected."

No further explanation is necessary.

---

§ 50.73(b)(2)(ii)(H)

The narrative description must include the following: "For failure that rendered a train of a safety system inoperable, an estimate of the elapsed time from the discovery of the failure until the train was returned to service."

No further explanation is necessary.

---

§ 50.73(b)(2)(ii)(I)

The narrative description must include the following: "The method of discovery of each component or system failure or procedural error."

Explain how each component failure, system failure, personnel error, or procedural deficiency was discovered. Examples include reviewing surveillance procedures or the results of surveillance tests, prestartup valve lineup check, performing quarterly maintenance, plant walkdown, and so forth.

---

§ 50.73(b)(2)(ii)(J)

The narrative description must include the following: "For each human performance related root cause, the licensee shall discuss the cause(s) and circumstances."

Generally, the criteria of 10 CFR 50.73(b)(2)(i) require a clear, specific narrative so that knowledgeable readers can understand the complete event. Further, for each human performance-related root cause, the criterion of 10 CFR 50.73(b)(2)(ii)(J) requires a description of the cause(s) and circumstances. In order to support an understanding of human performance issues related to the event, the narrative should address the factors discussed below to the extent they apply.

(1)   the cause(s), including any relation to the following areas:

(a)   procedures, where errors may be due to missing procedures, procedures that are inadequate due to technical or human factors deficiencies, or that have not been maintained current

(b)   training, where errors may be the result of a failure to provide training, having provided inadequate training, or as the result of training (such as simulator training or on-the-job training) that does not provide an environment comparable to that in the plant

(c)   communications, where errors may be due to inadequate, untimely, misunderstood, or missing communication or be due to the quality of the communication equipment

(d) human-system interface, such as the size, shape, location, function, or content of displays, controls, equipment, or labels, as well as environmental issues such as lighting, temperature, noise, radiation, and work area layout

(e) supervision and oversight, where errors may be the result of inadequate command and control, work control, corrective actions, self-evaluation, staffing, task allocation, overtime, or schedule design

(f) fitness for duty, where errors may be due to the influence of any substance legal or illegal, or mental or physical impairment; e.g., mental stress, fatigue, or illness

(g) work practices such as briefings, logs, work packages, teamwork, decisionmaking, housekeeping, verification, awareness, or attention

(2) the circumstances, including the following:

(a) the personnel involved, whether they are contractor or utility personnel, whether or not they are licensed, and the department for which they work

(b) the work activity being performed and whether or not there were any time or situational pressures present

---

§ 50.73(b)(2)(ii)(L)

The narrative description must include the following: "The manufacturer and model number (or other identification) of each component that failed during the event."

The manufacturer and model number (or other identification, such as type, size, or manufacture date) also should be given for each component found failed during the course of the event. An example of other identification could be (for a pipe rupture) size, schedule, or material composition.

(3) Safety System Responses

---

§ 50.73(b)(2)(ii)(K)

The narrative description must include the following: "Automatically and manually initiated safety system responses."

The LER should include a discussion of each specific system that actuated or failed to actuate. Do not limit the discussion to engineered safety features. Indicate whether or not the equipment operated successfully. For some systems, such as HPCI, RCIC, RHR, and AFW, the type of actuation may not be obvious. In those cases, indicate the specific equipment that actuated or should have actuated, by train, compatible with EPIX train definitions (e.g., AFW train B). Indicate the mode of operation, such as injecting into the reactor vessel, recirculation, pressure control, and any subsequent mode of operation during the event.

5.2.3 Assessment of Safety Consequences

> ### § 50.73(b)(3)
>
> The LER shall contain the following: "An assessment of the safety consequences and implications of the event. This assessment must include:
>
> (i) The availability of systems or components that could have performed the same function as the components and systems that failed during the event, and
>
> (ii) For events that occurred when the reactor was shutdown, the availability of systems or components that are needed to shutdown the reactor and maintain safe shutdown conditions, remove residual heat, control the release of radioactive material, or mitigate the consequences of an accident."

Give a summary assessment of the actual and potential safety consequences and implications of the event, including the basis for submitting the report. Evaluate the event to the extent necessary to fully assess the safety consequences and safety margins associated with the event.

Include an assessment of the event under alternative conditions if the incident would have been more severe (e.g., the plant would have been in a condition not analyzed in its latest FSAR) under reasonable and credible alternative conditions, such as a different operating mode. For example, if an event occurred while the plant was at low power and the same event could have occurred at full power, which would have resulted in considerably more serious consequences, this alternative condition should be assessed and the consequences reported.

Reasonable and credible alternative conditions may include normal plant operating conditions, potential accident conditions, or additional component failures, depending on the event. Normal alternative operating conditions and off-normal conditions expected to occur during the life of the plant should be considered. The intent of this section is to obtain the result of the considerations that are typical in the conduct of routine operations, such as event reviews, not to require extraordinary studies.

For events that occurred when the reactor was shut down, discuss the availability of systems or components that are needed to shut down the reactor and maintain safe shutdown conditions, remove residual heat, control the release of radioactive material, or mitigate the consequences of an accident.

5.2.4 Corrective Actions

> ### § 50.73(b)(4)
>
> The LER shall contain the following: "A description of any corrective actions planned as a result of the event, including those to reduce the probability of similar events occurring in the future."

Include whether the corrective action was or is planned to be implemented. Discuss repair or replacement actions as well as actions that will reduce the probability of a similar event occurring in the future. For example: "the pump was repaired and a discussion of the event was included in the training lectures." Another example: "although no modification to the instrument was deemed necessary, a caution note was placed in the calibration procedure for the instrument before the step in which the event was initiated."

In addition to a description of any corrective actions planned as a result of the event, describe corrective actions on similar or related components that were done, or are planned, as a direct result of the event. For example, if pump 1 failed during an event and required corrective maintenance and that same maintenance also was done on pump 2, so state.

If a study was conducted, and results are not available within the 60-day period, report the results of the study in a revised LER if they result in substantial changes in the corrective action planned. (See Section 5.1.5 for further discussion of submitting revised LERs.)

5.2.5 Previous Occurrences

---

§ 50.73(b)(5)

The LER shall contain the following: "Reference to any previous similar events at the same plant that are known to the licensee."

The term "previous occurrences" should include previous events or conditions that involved the same underlying concern or reason as this event, such as the same root cause, failure, or sequence of events. For infrequent events such as fires, a rather broad interpretation should be used (e.g., all fires and, certainly, all fires in the same building should be considered previous occurrences). For more frequent events, such as engineered safety feature actuations, a narrower definition may be used (e.g., only those scrams with the same root cause). The intent of the rule is to identify generic or recurring problems.

The licensee should use engineering judgment to decide how far back in time to go to present a reasonably complete picture of the current problem. The intent is to be able to see a pattern in recurring events, rather than to get a complete 10- or 20-year history of the system. If the event was a high-frequency type of event, 2 years back may be more than sufficient.

Include the LER number(s), if any, of previous similar events. Previous similar events are not necessarily limited to events reported in LERs. If no previous similar events are known, so state. If any earlier events, in retrospect, were significant in relation to the subject event, discuss why prior corrective action did not prevent recurrence.

5.2.6 Abstract (NRC Form 366, Item 16)

---

§ 50.73(b)(1)

The LER shall contain the following: "A brief abstract describing the major occurrences during the event, including all component or system failures that contributed to the event and significant corrective action taken or planned to prevent

---

recurrence."

Provide a brief abstract describing the major occurrences during the event, including all actual component or system failures that contributed to the event, all relevant operator errors or violations of procedures, the root cause(s) of the major occurrence(s), and the corrective action taken or planned for each root cause. If space does not permit describing failures, at least indicate whether or not failures occurred. Limit the abstract to 1,400 characters (including spaces), which is approximately 15 lines of single-spaced typewritten text. Do not use EIIS component function identifiers or the two-letter codes for system names in the abstract.

The abstract is typically included in the LER database to give users a brief description of the event to identify events of interest. Therefore, if space permits, provide the numbers of other LERs that reference similar events in the abstract.

As noted in Section 5.1.8, do not include safeguards, security, or proprietary information in the abstract.

5.2.7 Other Fields on the LER Form

(1) Facility Name (NRC Form 366, Item 1)

Enter the name of the facility (e.g., Indian Point, Unit 1) at which the event occurred. If the event involved more than one unit at a station, enter the name of the nuclear facility with the lowest nuclear unit number (e.g., Three Mile Island, Unit 1).

(2) Docket Number (NRC Form 366, Item 2)

Enter the docket number (in 8-digit format) assigned to the unit. For example, the docket number for Yankee-Rowe is 05000029. Note the use of zeros in this example.

(3) Page Number (NRC Form 366, Item 3)

Enter the total number of pages included (including figures and tables that are attached to Item 17 text) in the LER package. For continuation sheets, number the pages consecutively beginning with page 2. The LER form, including the abstract and other data, is prenumbered on the form as page 1 of _.

(4) Title (NRC Form 366, Item 4)

The title should include a concise description of the principal problem or issue associated with the event, the root cause, the result (why the event was required to be reported), and the link between them, if possible. It is often easier to form the title after writing the assessment of the event because the information is clearly at hand.

"Licensee Event Report" should not be used as a title. The title "Reactor Trip" is considered inadequate, because the root cause and the link between the root cause and the result are missing. The title "Personnel Error Causes Reactor Trip" is considered inadequate because of the innumerable ways in which a person could cause a reactor trip. "Technician Inadvertently Injected Signal Resulting in a Reactor Trip" would be a better title.

(5) Event Date (NRC Form 366, Item 5)

Enter the date on which the event occurred in the eight spaces provided. There are two spaces for the month, two for the day, and four for the year, in that order. Use leading zeros in the first and third spaces when appropriate. For example, June 1, 1987, would be properly entered as 06011987.

If the date on which the event occurred cannot be clearly defined, use the discovery date. See Section 2.5 of this report for further discussion of discovery date.

(6) Report Number (NRC Form 366, Item 6)

The LER number consists of three parts: (1) the four digits of the event year (based on event date), (2) the sequential report number, and (3) a revision number. The numbering system is shown in the diagram below; the event occurred in the year 1991, it was the 45th event of that year, and the submittal was the first revision to the original LER for that event.

Event Year	Sequential Report Number	Revision Number
1991 -	045 -	01

Event Year: Enter the four digits. The event year should be based on the event date (Item 4).

Sequential Report Number: As each reportable event is reported for a unit during the year, it is assigned a sequential number. For example, for the 15th and 33rd events to be reported in a given year at a given unit, enter 015 and 033, respectively, in the spaces provided. Follow the guidelines below to ensure consistency in the sequential numbering of reports.

- Each unit should have its own set of sequential report numbers. Units at multiunit sites should not share a set of sequential report numbers.

- The sequential number should begin with 001 for the first event that occurred in each calendar year, using leading zeros for sequential numbers less than 100.

- For an event common to all units of a multiunit site, assign the sequential number to the lowest numbered nuclear unit.

- If a sequential number was assigned to an event, and it was subsequently determined that the event was not reportable, a "hole" in the series of LER numbers would result. The NRC would prefer that licensees reuse a sequential number rather than leave holes in the sequence. A sequential LER number may be reused even if the event date was later than subsequent reports.

If the licensee chooses not to reuse the number, write a brief letter to the NRC noting that "LER number xxx for docket 005000XXX will not be used."

Revision Number: The revision number of the original LER submitted is 00. The revision number for the first revision submitted should be 01. Subsequent revisions should be numbered sequentially (i.e., 02, 03, 04).

(7) Report Date (NRC Form 366, Item 7)

Enter the date the LER is submitted to the NRC in the eight spaces provided, as described in Section 5.2.7(5) above.

(8) Other Facilities (NRC Form 366, Item 8)

When a situation is discovered at one unit of a facility that applies to more than the one unit, submit a single LER. LER form Items 1, 2, 6, 9, and 10 should refer to the unit primarily affected or, if both units were affected approximately equally, to the lowest numbered nuclear unit.

The intent of the requirement is to name the facility in which the primary event occurred, whether or not that facility is the lowest numbered of the facilities involved. The automatic use of the lowest number should only apply to cases where both units are affected approximately equally. Item 8 only should indicate the other unit(s) affected. The abstract and the text should describe how the event affected all units.

Enter the facility name and unit number and docket number (see Sections 5.2.7(1) and 5.2.7(2) for format) of any other units at that site that were directly affected by the event (e.g., the event included shared components, the LER described a tornado that threatened both units of a two-unit plant).

(9) Operating Mode (NRC Form 366, Item 9)

Enter the operating mode of the unit at the time of the event as defined in the plant's TS in the single space provided. For plants that have operating modes such as hot shutdown, cold shutdown, and operating but do not have numerical operating modes (e.g., Mode 5), place the letter "N" in Item 9 and describe the operating mode in the text.

(10) Power Level (NRC Form 366, Item 10)

Enter the percent of licensed thermal power at which the reactor was operating when the event occurred. For shutdown conditions, enter 000. For all other operating conditions, enter the correct numerical value (estimate power level if it is not known precisely), using leading zeros as appropriate (e.g., 009 for 9-percent power). Significant deviations in the operating power in the balance of plant should be clarified in the text.

(11) Reporting Requirements (NRC Form 366, Item 11)

Check one or more blocks according to the reporting requirements that apply to the event. A single event can meet more than one reporting criterion. For example, if, as a result of sabotage, reportable under 10 CFR 73.71(b), a safety system failed to function, reportable under 10 CFR 50.73(a)(2)(v), and the net result was a release of radioactive material in a restricted area that exceeded the applicable license limit, reportable under 10 CFR 20.2203(a)(3)(i), prepare a single LER and check the three boxes for 10 CFR 73.71(b), 50.73(a)(2)(v), and 20.2203(a)(3)(i).

In addition, an event can be reportable as an LER even if it does not meet any of the criteria of 10 CFR 50.73. For example, a case of attempted sabotage (10 CFR 73.71(b)) that does not result in any consequences that meet the criteria in 10 CFR 50.73 can be reported using the

"Other" block. Use the "Other" block if a reporting requirement other than those specified in Item 11 was met. Specifically describe this other reporting requirement in the space provided below the "Other" block and in the abstract and text.

(12) <u>Licensee Contact (NRC Form 366, Item 12)</u>

---

§ 50.73(b)(6)

The LER shall contain the following: "The name and telephone number of a person within the licensee's organization who is knowledgeable about the event and can provide additional information concerning the event and the plant's characteristics."

Enter the name, position title, and work telephone number (including area code) of a person who can provide additional information and clarification for the event described in the LER.

(13) <u>Component Failures (NRC Form 366, Item 13)</u>

Enter the appropriate data for each component failure described in the event. A failure is defined as the termination of the ability of a component to perform its required function. Unannounced failures are not detected until the next test; announced failures are detected by any number of methods at the instant of occurrence.

If multiple components of the same type failed and all of the information required in Item 13 (i.e., cause, system, component, and so forth) was the same for each component, then only a single entry is required in Item 13. Clearly define the number of components that failed in the abstract and text.

The component information elements of this item are discussed below.

<u>Cause</u>: Enter the cause code as shown below. If more than one cause code is applicable, enter the cause code that most closely describes the root cause of the failure.

Cause
Code   <u>Classification and Definition</u>

A   <u>Personnel Error</u> is assigned to failures attributed to human errors. Classify errors made because written procedures were not followed or because personnel did not perform in accordance with accepted or approved practice as personnel errors. Do not include errors made as a result of following incorrect written procedures in this classification.

B   <u>Design, Manufacturing, Construction/Installation</u> is assigned to failures reasonably attributed to the design, manufacture, construction, or installation of a system, component, or structure. For example, include failures that were traced to defective materials or components otherwise unable to meet the specified functional requirements or performance specifications in this classification.

C   <u>External Cause</u> is assigned to failures attributed to natural phenomena. A typical example would be a failure resulting from a lightning strike, tornado, or flood. Also assign this classification to manmade external causes that originate off site (e.g., an industrial accident at a nearby industrial facility).

D   Defective Procedure is assigned to failures caused by inadequate or incomplete written procedures or instructions.

E   Management/Quality Assurance Deficiency is assigned to failures caused by inadequate management oversight or management systems (e.g., major breakdowns in the licensee's administrative controls, preventive maintenance program, surveillance program, or quality assurance controls; inadequate root cause determination; inadequate corrective action).

X   Other is assigned to failures for which the proximate cause cannot be identified or which cannot be assigned to one of the other classifications.

System: Enter the two-letter system code from Institute of Electrical and Electronics Engineers (IEEE) Standard 805-1984, "IEEE Recommended Practice for System Identification in Nuclear Power Plants and Related Facilities," dated March 27, 1984. Copies may be obtained from the Institute of Electrical and Electronics Engineers, 445 Hoes Lane, P.O. Box 1331, Piscataway, NJ 08855-1331.

Component: Enter the applicable component code from IEEE Standard 803A-1983, "IEEE Recommended Practice for Unique Identification in Power Plants and Related Facilities—Component Function Identifiers."

Component Manufacturer: Enter the four-character alphanumeric reference code. If the manufacturer is one used in EPIX, use the manufacturer's name as it appears in EPIX.

Reportable to EPIX: Enter a "Y" if the failure is reportable to EPIX and an "N" if it is not reportable.

Include in the LER text and in Item 13 of the LER form any component failure involved in the event, not just components within the scope of EPIX or EIIS.

Failure Continuation Sheet (NRC Form 366B): If more than four failures need to be coded, use one or more of the failure continuation sheets (NRC Form 366B). Code the entries in Items 1, 2, 3, and 6 of the failure continuation sheet to match entries of these items on the initial page of the LER. Complete Item 13 in the same manner as Item 13 on the basic LER form. Do not repeat failures coded on the basic LER form on the failure continuation sheet. Place any failure continuation sheets after any text continuation sheets and include those sheets in the total number of pages for the LER.

(14) Supplemental Report (NRC Form 366, Item 14)

Check the "Yes" block if the licensee plans to submit a followup report. For example, if a failed component had been returned to the manufacturer for additional testing and the results of the test were not yet available when the LER was submitted, a followup report would be submitted.

(15) Expected Submission Date of Supplemental Report (NRC Form 366, Item 15)

Enter the expected date of submission of the supplemental LER, if applicable. See Section 5.2.7(5) for the proper date format. The expected submission date is a target/planning date; it is not a regulatory commitment.

(16) LER Text Continuation Sheet (NRC Form 366A)

Use one or more additional text continuation sheets of LER Form 366A to continue the narrative, if necessary. There is no limit on the number of continuation sheets that may be included.

Drawings, figures, tables, photographs, and other aids may be included with the narrative to help readers understand the event. If possible, provide the aids on the LER form (i.e., NRC Form 366A). In addition, care should be taken to ensure that drawings and photographs are of sufficient quality to permit legible reproduction and micrographic processing. Avoid oversized drawings (i.e., larger than 8.5 by 11 inches).

5.2.8 Obtaining LER Forms

The latest NRC forms may be found at the NRC public Web site at http://www.nrc.gov/reading-rm/doc-collections/forms/.

NRC FORM 335 (12-2010) NRCMD 3.7	U.S. NUCLEAR REGULATORY COMMISSION	1. REPORT NUMBER (Assigned by NRC, Add Vol., Supp., Rev., and Addendum Numbers, If any.)
	BIBLIOGRAPHIC DATA SHEET *(See instructions on the reverse)*	NUREG-1022, Rev. 3

2. TITLE AND SUBTITLE	3. DATE REPORT PUBLISHED	
Event Reporting Guidelines: 10 CFR 50.72 and 50.73 Final Report	MONTH	YEAR
	January	2013
	4. FIN OR GRANT NUMBER	

5. AUTHOR(S)	6. TYPE OF REPORT
	Technical
	7. PERIOD COVERED (Inclusive Dates)

8. PERFORMING ORGANIZATION - NAME AND ADDRESS (If NRC, provide Division, Office or Region, U. S. Nuclear Regulatory Commission, and mailing address; if contractor, provide name and mailing address.)
Division of Inspection and Regional Support
Office of Nuclear Reactor Regulation
U.S. Nuclear Regulatory Commission
Washington D.C. 20555-0001

9. SPONSORING ORGANIZATION - NAME AND ADDRESS (If NRC, type "Same as above", if contractor, provide NRC Division, Office or Region, U. S. Nuclear Regulatory Commission, and mailing address.)
Same as above.

10. SUPPLEMENTARY NOTES
A. Lewin, NRC Project Manager

11. ABSTRACT (200 words or less)
NUREG-1022, "Event Reporting Guidelines: 10 CFR 50.72 and 50.73," contains guidelines that the staff of the U.S. Nuclear Regulatory Commission (NRC) considers acceptable for use in meeting the requirements of Title 10 of the Code of Federal Regulations (10 CFR) 50.72 and 50.73. Several identified reporting issues could not be quickly resolved given certain ambiguities in NUREG-1022, Revision 2 guidance. In developing Revision 3 to NUREG-1022, the NRC held numerous public and internal meetings to solicit stakeholder inputs and feedback. In resolving the ambiguities, the NRC considered the provisions of the rule itself, the associated statements of consideration, and other available guidance in that hierarchal order. Revision 3 to NUREG-1022 revises the event reporting guidelines in order to provide clearer guidance.

12. KEY WORDS/DESCRIPTORS (List words or phrases that will assist researchers in locating the report.)	13. AVAILABILITY STATEMENT
NUREG-1022, Event Reporting Guidelines, 10 CFR 50.72, 10 CFR 50.73	unlimited
	14. SECURITY CLASSIFICATION
	(This Page) unclassified
	(This Report) unclassified
	15. NUMBER OF PAGES
	16. PRICE

NRC FORM 335 (12-2010)

UNITED STATES
NUCLEAR REGULATORY COMMISSION
WASHINGTON, DC 20555-0001

OFFICIAL BUSINESS

NUREG-1022, Rev. 3
Final

Event Report Guidelines: 10 CFR 50.72 and 50.73

January 2013